SC_ Secret Operations Manual

SOE Secret Operations Manual

SOE: Secret Operations Manual

Copyright © 1993 by Paladin Press

ISBN 13: 978-0-87364-744-1

Printed in the United States of America

Published by Paladin Press, a division of
Paladin Enterprises, Inc.
Gunbarrel Tech Center
7077 Winchester Circle
Boulder, Colorado 80301 USA
+1.303.443.7250

Direct inquiries and/or orders to the above address.

PALADIN, PALADIN PRESS, and the "horse head" design are trademarks belonging to Paladin Enterprises and registered in United States Patent and Trademark Office.

All rights reserved. Except for use in a review, no portion of this book may be reproduced, stored in or introduced into a retrieval system, or transmitted in any form without the express written permission of the publisher. The scanning, uploading and distribution of this book through the Internet or any other means without the permission of the publisher is illegal and punishable by law. Please respect the author's rights and do not participate in the any form of electronic piracy of copyrighted material.

Neither the author nor the publisher assumes any responsibility for the use or misuse of information contained in this book.

Visit our website at www.paladin-press.com

INDEX

DOUBLE TRANSPOSITION CIPHER	A
ANONYMOUS LETTERS IN CONNECTION WITH PASSIVE RESISTANCE AND SIMPLE SABOTAGE	B
PROPAGANDA - INTRODUCTION	C
CIPHER, Elementary Course	D
UNDERGROUND IN THE FAR EAST	E
PROPAGANDA PRESENTATION	F
OPINION SAMPLING	G
SELECTION OF TARGETS	H
PLANNING AND METHODS OF ATTACK	I
PASSIVE RESISTANCE	J
LIAISON AGENT	K
INTERROGATION	L
BURGLARY	M

A DOUBLE TRANSPOSITION CIPHER

SECRET

Why we use cipher

Our messages frequently have to be sent through channels that are not physically secure from the enemy. Radio signals, for example, are open to anyone with a receiving set. Our enemy, as we do, maintains monitoring stations that record almost everything on the air.

To try to prevent the enemy from reading our messages sent by radio and other open channels, we put them into code or cipher. Each message is an incoherent series of letters, such as

TATON HNRSI EAISA NTISS ISYPO

This jumble of letters is meaningless to the casual reader, but it actually can be resolved into coherent words by anyone having knowledge of the key.

Our system

The cipher system here described is <u>two-phase</u> or <u>double transposition</u>. Every feature of it is known to the enemy, who, indeed, also uses it. It is the system most feasible for use by undercover agents. The only point of secrecy, the variable that the enemy does not know, is the <u>specific key</u> of a particular message.

Keys

To encipher any message, we use two <u>numerical keys</u>. By a numerical key we mean a mixed sequence of integers from 1 up, as

12 9 1 4 10 8 6 11 2 13 3 7 5

SECRET

To make the system sufficiently secure, we must never transmit two messages enciphered under the same keys. Consequently, when an agent leaves his base he is provided with at 100 pairs of keys. He numbers his message serially from 1 to 100, and the numbers correspond with key-pairs on a written list held by his base. The agent himself cannot risk carrying with him any written memorandum or ciphering device that would betray his use of cipher. Therefore he has to <u>memorize</u> the list.

You may well ask, how can a man memorize 200 different permutations of integers? The answer is, no one can be expected to do that. What the agent actually memorizes is <u>words</u>. It is no trick at all to memorize a few hundred words. The numerical keys are derived from work keys by a very simple method.

Derivation of numerical keys

You must first of all know the order of the letters in what we call the <u>normal</u> English alphabet:

A B C D E F G H I J K L M N O P Q R S T U V W X Y Z

Suppose that the given keyword is TRANSPOSITION. To derive the numerical key, first look for any A's in this word. There is an A; write under it number "1".

```
T   R   A   N   S   P   O   S   I   T   I   O   N
        1
```

Next look for B's; there are none. Similarly, look in turn for each letter, in the order of the normal alphabet. There is no C, D, E, F, or G. The next letters to be numbered are the two I's.

Whenever the same letter occurs two or more times, we number these occurrences in order from left to right:

```
T R A N S P O S I T I O N
1               2       3
```

Continue in the same way until a number has been written under every letter of the keyword. The complete result:

```
T  R  A  N  S  O  S  I  T  I  O  N
12 9  1  4  10 8  6  11 2  13 3  7  5
```

The keys we actually use are somewhat longer than this example. In order to find sequences long enough, we may use phrases as well as single words. If the "keyword" is actually several words, write them consecutively with no spaces between, thus

```
T  H  E  B  A  L  T  I  M  O  R  E  S  U  N
13 5  3  2  1  7  14 6  8  10 11 4  12 15 9
```

Exercise

Derive the numerical key from each of the following phrases:

1. THE WASHINGTON TIMES HERALD
2. AMERICAN TELEPHONE AND TELEGRAPH
3. PHILADELPHIA PENNSYLVANIA
4. UNDERWOOD NOISELESS TYPEWRITER
5. THE NEW YORK TIMES
6. CO-ORDINATOR OF INFORMATION
7. WEBSTER'S STANDARD DICTIONARY
8. DEPARTMENT OF AGRICULTURE
9. SECRETARY OF STATE
10. SAN FRANCISCO CALIFORNIA

Construction of Box 1

The two keys of a message are labelled Key 1 and Key 2, in the order in which they are used for encipherment.

To encipher a clear message, commence by

SECRET

writing Key 1 in a row (horizontal). Write the numerical key under it and draw a heavy line just below the numerical key. This line is the top of the <u>encipherment box</u>. The box will be just as wide as the key. Then draw vertical lines just to left and right of the key, from the key-rows downward. Simplest is to draw these lines a little shorter than you estimate they will have to be, and later lengthen them to fit.

The bottom of the box is at the outset left open, since you usually don't know how many letters you are going to write. Don't waste time counting the number of letters in your message. It will be much easier to count them after they have been written into Box 1.

Write your message into the box in <u>normal</u> manner. Start at the upper left corner, write in rows from left to right, and use the rows in order from top down. Write just as many letters in each row as there are letters in the key. This work is of course easier if you have paper ruled into squares or cells. Don't leave any cell empty in the course of writing the message. Close up the words solid.

Here is an example of a message written into Box 1:

```
W   A   S   H   I   N   G   T   O   N
10  1   8   3   4   5   2   9   7   6

J   A   P   A   N   E   S   E   A   D
V   A   N   C   E   S   I   N   T   H
E   P   H   I   L   I   P   P   I   N
E   S   W   E   R   E   H   U   R   L
E   D   B   A   C   K   Y   E   S   T
E   R   D   A   Y   B   Y   G   E   N
E   R   A   L   M   A   C   A   R   T
H   U   R
```

The foot

It is a regulation that every encipherment box must have a _foot_. By this term we mean an incomplete line at the end of the box. This irregularity in what is otherwise a rectangle increases the cryptographic security of the cipher.

The foot is most effective when it is about half as long as the key. You cannot always contrive to make it so, but you must be sure that the foot is neither too long nor too short. The regulation is that the foot must contain at least two cells, and not more than $n-2$ cells where n is the length of key.

Multiple of 5

In radio, telegraph or cable transmission, cipher letters are sent in groups of five. It is very undesirable to send a short group at the end of a message. Therefore the regulations require that the total number of letters in any DT cipher be a multiple of 5.

Nulls

If the number of letters we have written into Box 1 does not change to be a multiple of 5, we must make it so by adding sufficient _nulls_.

Nulls are letters that have no meaning, but are added or inserted merely to conform to regulations or to increase cryptographic security.

Here are some cautions about the choice of nulls:

(a) For nulls adjacent to plain text choose letters that obviously are not a part of it.
(b) Avoid falling into the habit of using

the same sequence of nulls in every message. Vary your style in nulls from one message to the next.

(c) Choose nulls predominantly (not necessarily wholly) from the English high-frequency letters: vowels A E I O U and consonants N R S T.

(d) If you have to add a large number of terminal nulls, you may write words instead of incoherent letters. Be careful that the words are clearly alien to the message.

(e) It is common practice (and a wise precaution) to commence with some initial nulls. The regulation is that no doubled letter may appear in the initial nulls until the plain text is about to begin; then the last null must be doubled to mark the break.

Closing up Box 1

We have written a message into Box 1. We are ready to close the bottom of the box. Four regulations have to be observed in so doing:

1. The total of letters must be a multiple of 5.
2. Box 1 must have a foot.
3. Box 2 must have a foot.
4. The bottom of the box must not lie in the danger zone.

The easiest way to proceed is to add groups of five to the total number of letters in the message. After adding the first group of five letters, take stock and see if this satisfies all the conditions above. If it does not, add another group of five letters, and take stock again. Do this until all the above points are satisfied.

To satisfy 2 and 3, the chosen number must not be an exact multiple of either key-length. For example, our message under key WASHINGTON (page 4) contains 73 letters. We can reach a multiple of 5 by adding two nulls. But the number 75 is unsuitable if our second key is 15 in length,

for then Box 2 would have no foot. If we add
instead seven nulls to reach 80, Box 2 has a foot
but Box 1 does not.

It is not enough to choose a total number
that will give a foot on each box, that is,
between 2 and n-2 letters. We must also assure
that each foot is of permissible size within the
regulations. The way to determine the dimensions
of a box is described below.

<u>Determining dimensions of a Box</u>

A cipher box consists of a rectangle plus a
foot. To determine the correct dimensions of a
box which is to contain a given number of letters,
divide this number by the number of letters in the
key. The quotient is the other dimension of the
rectangle (length at right angles to the key) and
the remainder is the number of cells in the foot.

For example, 75 letters are to be put into a
box under a key of 12 letters. The division,
75/12 gives quotient 6 and remainder 3. The depth
of the box (right-hand column) will be 6 lines,
and the foot will have 3 cells.

<u>Danger Zone</u>

As is explained later, the columns of letters
of Box 1 are copied into the columns of Box 2. For
cryptographic security it is of utmost importance
that no column of Box 1 shall exactly fit into the
column-length of Box 2. The incoming columns must
"break joints."

Box 1 contains columns of two different
lengths -- short (e.g. rightmost) and long (e.g.
leftmost). Neither length may be equal to, or an
integral multiple of, nor an integral divisor or,
the standard length of the Box 2 columns.

Since the standard length of Box 2 columns is
the same as the key-length, it is easy to stay out
of the danger zone in closing Box 1. Simply note
the length of Key 2 and void closing Box 1 at a
point that will create any column of a length that
will divide evenly into this number.

For example, suppose that Key 2 is 16 letters
long. If we close up the box on page 3 by adding
only two nulls, Box 1 has long columns of 8. But
8 is an integral divisor of 16. If we fill out
the 8th row of Box 1 with nulls and place the foot
on the 9th row, we will have columns of 8 on the
short side. Hence we will have to fill up the 9th
row also and place the foot not earlier than the
10th row.

Construction of Box 2

Write Key 2 in a column, from top down.
Write the numerical key in the next column at
right. Then turn the paper counter-clockwise 90
degrees and build Box 2 <u>above</u> the key.

The foot will thus be at the top of the box.
But directions as to right and left are the same
as for Box 1. The foot is at the left, and the
keyword reads from left to right.

Turn the paper back to the original position,
with the key word written in a vertical column,
before starting to write letters into it. Below
is shown Box 2 prepared for 75 letters.

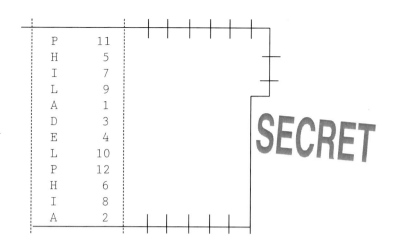

Transcription from Box 1 to Box 2

Read the letters <u>by columns</u> out of Box 1, taking the columns in order as given by the numerical key (start with column 1, then 2, and so on). Always read columns from the top down.

Write the letters into the <u>columns</u> of Box 2, taking these columns <u>in normal order</u>. That is, start at the upper left corner of Box 2 (next to the first letter of the keyword), fill the column from top down, next fill the column immediately to right of the first, and so on. The last part of the box to be filled is the foot.

To continue the sample encipherment:

Box 1

W	A	S	H	I	N	G	T	O	N
10	1	8	3	4	5	2	9	7	6
J	A	P	A	N	E	S	E	A	D
V	A	N	C	E	S	I	N	T	H
E	P	H	I	L	I	P	P	I	N
E	S	W	E	R	E	H	U	R	L
E	D	B	A	C	K	Y	E	S	T
E	R	D	A	Y	B	Y	G	E	N
E	R	A	L	M	A	C	A	R	T
H	U	R	D	N	:				

Box 2

P	11	A	Y	E	B	R	E	E
H	5	A	Y	L	A	S	N	E
I	7	P	C	R	D	E	P	H
L	9	S	A	C	H	R	U	—
A	1	D	C	Y	N	P	E	
D	3	R	I	M	L	N	G	
E	4	R	E	N	T	H	A	
L	10	U	A	E	N	W	J	
P	12	S	A	S	T	B	V	
H	6	I	L	I	A	D	E	
I	8	P	D	E	T	A	E	
A	2	H	N	K	I	R	E	

Box 1 was closed up with a total of 75 letters, a number suitable in every way to the length 12 of the second key. In transcription, column 1 of Box 1, AAPS..etc. was first copied into the leftmost column of Box 2. Then column 2, SIPH...etc. was taken out and were written into Box 2 in continuation of the normal order.

Taking out Cipher Text

The cipher text is taken from Box 2. Copy the letters from the <u>rows</u> of Box 2, reading left to right, and taking the rows in order of the numerical key (from 1 up).

Write the cipher text in rows from left to right, clear across your paper, just as you do in normal message-writing. Group the letters into fives, leaving one space between each group. The cipher text derived from the above example is:

DCYNP EHNKI RERIM LNGRE NTHAA YLASN EILIA DEPCR

DEPHP DETAE SACHR UUAEN WIAYE* BREES ASTBV

(*This pentagraph is really WJAYE, but in accordance with regulations the J has been changed to I. Don't forget that the letters J, Q, Z are not to be written in cipher texts. Change these letters respectively to I, C, S.)

Indicators

Before the cipher text is ready for transmission, you must add <u>indicators</u> to it. The indicators are very important, for without them your correspondent will not know how to decipher your message.

The indicator is a group of five numerals. The first two digits of the group give the serial number of the message. Since you must write this number in two digits, no more and no less, write 01 for your first message, 02 for your second, and so on. After 99, your next message is 00, meaning 100. Then you start over again at 01 with a new set of keys.

The last three digits of the indicator give the total number of letters in the cipher text

(or, what is the same thing, in either Box 1 or
Box 2.) For example, if our message JAPANESE
ADVANCES etc., is our first report to base, the
indicator will be 01075.

The indicator group is written twice, once
before the first group of the cipher text, and
again after the last group. Thus:

01075 DCYNP EHNKI RERIM LNGRE NTHAA YLASN EILIA DEPCR

DEPHP DETAE SACHR UUAEN WIAYE BREES ASTBV 01075

<u>Exercises</u>

Encipher each of the following messages,
using the given keys. Be sure to add appropriate
indicators.

1. DAILY REPORT NOT YET RECEIVED X PLEASE SEND IN
   ALL AVAILABLE INFORMATION IMMEDIATELY

    Key 1: GETTYSBURG   Key 2: PENNSYLVANIA

2. FIVE MORE JAPANESE TRANSPORTS HAVE BEEN SUNK BY
   CRUISERS AND DESTROYERS OF THE ASIATIC FLEET
   TODAY

    Key 1: INSTRUCTIONS   Key 2: OFFICIAL

3. BRITISH TROOPS FOUGHT IN A STEADY DOWNPOUR AND
   BOOT HIGH MUD YESTERDAY

    Key 1: FUNDAMENTAL   Key 2: SAFEGUARDS

<u>Decipherment</u>

Once you have learned encipherment, you can
figure out for yourself how to perform decipherment. Just review in your mind the steps taken by
the encipherer to make the cipher text, then undo
or reverse these steps in turn, commencing with

the last and ending with the first.   SECRET

Here is a summary of the routine of decipherment:

1. Count the number of letters in the cipher text to see whether it corresponds with the indicator. Don't overlook this precaution, because sometimes you may not receive all of the cipher text.

2. From the serial number of the message ascertain the two keys.

3. Write the two keywords, derive the numerical keys, and design the two boxes just as for encipherment. Since you know the number of letters to be put into each box, it can be outlined complete.

4. Copy the cipher text into the <u>rows</u> of Box 2, filling first row 1, then row 2, and so on, in accordance with the key. Of course write from left to right.

5. Having filled Box 2 correctly, next transcribe the letters to Box 1. Read them out of the <u>columns</u> of Box 2 <u>in normal order</u>. That is, start at the upper left corner, read down the column immediately to the right, and so on, ending with the foot. Write the letters as you read them into the <u>columns</u> of Box 1, filling first column 1, then column 2, and so on, in accordance with the key.

6. Having filled Box 1 correctly, you can read the plain text in normal manner -- on rows from left to right and from top row downward.

<u>Horizontal or vertical?</u>

If at any point in enciphering or deciphering you become confused, and are uncertain which way

to read or which way to write, stop work and
review the operations of encipherment.

Remember that the first step in this direct
process is to write a message <u>normally</u>. The plain
text is written just as you would write a letter --
in rows from left to right and top downward.

The rest is a matter of commonsense. Letters
must be taken out of a box in the direction opposite from that in which they were put in, else
there would be no transposition. That means they
must be taken out of Box 1 by columns.

The layout is arranged so that you <u>write</u> letters in the same direction as you <u>read</u> them.
Hence the letters read out of Box 1, must be written by columns into Box 2. To make the second
transposition, they must be taken out again in the
opposite way, horizontally. That means that the
rows of Box 2 are copied out to make the rows of
the cipher text.

To clarify your mind on any step of decipherment, review encipherment and think what you have
to do in order to undo the operations of the inci-
pherer.

Read the following observations. Any one of
them, if you remember, may prove a shortcut toward
clearing up uncertainty.

(a) The relation between the two boxes, in
either encipherment or decipherment, is always
<u>columnar</u>.

(b) The sequence of directions (vertical or
horizontal) is the same for both encipherment and
decipherment. In either case there are three
steps, as follows:

HORIZONTAL -- (write plain text into Box 1 or

cipher text into Box 2)
     VERTICAL -- (transcribe letters from Box 1 to Box 2 or vice versa)
     HORIZONTAL -- (copy cipher text out of Box 2 or read plain text out of Box 1)

     (c) In dealing with <u>numbered</u> lines (i.e., lines at right angles to the key) follow the numbers of the key. In dealing with <u>unnumbered</u> lines (parallel to the key) follow normal order (left to right and top to bottom for rows; top to bottom and left to right for columns). These rules apply both to reading letters out and writing letters in.

<u>Exercises</u>

     Decipher the following messages, using the given keys.

1. 01085 RLONW LDETH TEWSP IPTBI IMGIU ELGRT IDAES TNVHS AAFAE WASLS MRAFH DTASP ESTKE SLONE OBYNN HEWNE UEPAA 01085

Key 1:   MILITARY INTELLIGENCE
Key 2:   DANGEROUS MISSION

2. 02115 FLYPT PPETY LENBR LILOF OEMEN SHTTT TAYRG ALEHA XPTNL TOAYI AOHSH KFEAE EAFTM AESTN ATTEN OTDRE OLNCA TNIEV NHRAN HIWOL SNMRU OEVFU NYOMM 02115

Key 1:   CONSTITUTION AVENUE
Key 2:   CONFEDERATE STATES

3. 03115 AEAHR NWART YRSNN PIOGI EETRC IEMTD BNITR RNAOT FOFWE ONOUT OETYO NCARO CIROS KTVAO ERTPE NEENO RIKOH YFTPI ALOIM LRIOT TAGFW NBRAV DEDAP 03115

Key 1:   GOD BLESS AMERICA
Key 2:   MARY HAD A LITTLE LAMB

B

ANONYMOUS LETTERS IN CONNECTION WITH PASSIVE RESISTANCE AND SIMPLE SABOTAGE.

AN ANONYMOUS LETTER IS A WEAPON which is readily available to any literate person. No special materials are required except those which are accessible to all persons. By using reasonable care a maximum of security is afforded. The anonymous letter writer and the amount of damage which can be accomplished amongst enemy personnel is only limited by the ingenuity of the writer applied to the environment in which he is living.

<u>OBJECTIVES</u>

1. Creating general distrust and suspicion.

    a). It is a rule of human behavior that the receipt of an anonymous letter will cause distrust and suspicion even though the recipient pretends that he is too "high minded" to give such a document any consideration. Standard policy procedure involves at least a routine investigation of allegations made in all anonymous letters. Even indiscriminate circulation of anonymous letters with accusations therein picked at random may occasionally expose condition which the victim would prefer to keep hidden.

2. An anonymous letter is a ready means of discrediting your enemy through the exposure of an actual condition of which you have authentic knowledge.

3. An anonymous letter may serve as a spark to initiate a "frame up" of hostile

officials after evidence has previously been planted. Don't be too accurate. Fabricate a bit.

4. Because all anonymous letters must be carefully investigated, constant circulation of such missives will overburden police and counter espionage organizations with useless investigations with a resultant waste in manpower and time. For example, an anonymous "tip" that a railway junction was to be the object of immediate sabotage would cause a considerable useless diversion of enemy forces.

5. A heavy circulation of anonymous letters which have no authentic foundation will in itself tend to protect the subversive agent against similar denunciations directed against himself. Constant useless investigation of such letters preconditions enemy forces to expect negative results and also gives our agent a ready explanation for such an accusation.

6. An anonymous letter carefully directed can be cause of serious material discord amongst enemy families with a possible resultant loss of morale and efficiency.

MEDIUM

1. Postal letter.
2. Planted note.
   a). Note planted for a particular recipient.
   b). Note planted for random recipient.
3. Letters to newspapers.
   a). Since newspapers will not generally print unsigned or scandalous communications, considerable subtlety must be employed in using this medium.

4. Handbills.
5. Lettering in public places.

SECURITY

Avoid duplicate style, except where contents are authentic, where your accurate knowledge of more than one situation would lead to identification.

MECHANICAL MEDIUM

1. Handwriting.

    a). Printing.
    b). Script.

2. Instruments.

    a). Pencil.
    b). Pen.
    c). Typewriter.
    d). Rubber stamp.
    e). Chalk or crayon.
    f). Letters cut from printed matter.

3. Paper or background.

    a). Avoid distinctive paper.
    b). Avoid watermarks.
    c). Avoid any paper of which you have quantities in your own possession unless such possession is common to most residents of that area.

4. <u>Avoid distinctive spelling or mis-spelling.</u>

5. Grammar.

6. Idioms.

7. Slang.

8. Postmark.

9. Address.

DIRECTION

1. To general public.

    a). Newspapers.
    b). Handbills.
    c). Lettered signs in public places.

2. To wives, husbands, sweethearts, parents.

3. To police and military authorities.

4. To civil officials.

5. To business associates.

CONTENTS

1. Accusations.

    a). Disloyalty.
    b). Civil criminal acts.

        1. Official corruption.
        2. Theft.
        3. Fraud.
        4. Sex crimes.
        5. Hoarding.
        6. Smuggling.
        7. Drug addiction.
        8. Illegal money transactions.

    c). Sexual misconduct.
        1. Infidelity.
        2. Perversion.

2. Threats of:

   a). Death, injury or abduction to self or family.
   b). Destruction of personal property.
   c). Destruction of public property.

3. "Warnings".

   a). Of death, etc., to officials or relatives.
   b). Of destruction of public or private property.

C

PROPAGANDA - INTRODUCTORY

1. <u>WHAT IS PROPAGANDA?</u> **SECRET**

   There are two kinds of propaganda.

   a) <u>Preparational</u>.

   "The art of persuasion with a view to producing merely, a frame of mind".

   E.g. Goebbels persuaded German people to adopt a mental attitude vis-a-vis the concept "Lebensraum" before he called on them to <u>do</u> anything about it.

   b) <u>Operational</u>.

   "The art of persuasion with a view to producing action."

   E.g. Today it is useless for our propaganda merely to persuade Frenchmen that the Boche is a swine. It must also instruct Frenchmen how to kick the Boche out of France.

   Our propaganda to enemy and occupied countries is now mainly operational; and as such should always contain in the joint elements of persuasion and action.

2. <u>WHAT CAN PROPAGANDA DO</u>?

   These lectures deal only with underground operational propaganda - i.e. with propaganda as one weapon in the whole armory of underground warfare. Passive resistance, sabotage, guerrilla warfare and internal revolution are other weapons which, with propaganda, must be knitted into a whole.

Thus propaganda calling in its action elements for passive resistance may lead to passive resistance; passive resistance plus propaganda may lead to sabotage; passive resistance and sabotage plus propaganda to guerrilla warfare, etc. (Cf. oil in a machine.)

Therefore propaganda, though an important weapon, is never an independent one. It must be co-ordinated with the other weapons at our disposal.
(Cf. Goebbels "Fourth Arm.")

3. WHY DO YOU NEED TO KNOW ABOUT PROPAGANDA?

   If ever you are called on to become an organizer, propaganda will be one of the weapons at your disposal. It is therefore right that you should know its scope, its power, and its relation to the conduct of underground warfare.

   We shall best perceive these things if we follow, in some detail, the career of a propaganda-agent from the moment when, in this country, he is told to undertake propaganda work to the moment when, in his own country, he has distributed his first leaflet among his people.

4. STARTING POINT.

   The propagandist finds his raw material in the facts of the political situation. A thorough knowledge of such facts as they affect his field of operations is essential for successful work.

   SECRET

5. POLITICS.

   The propagandist approach to politics should be governed by two main considerations:

a) Politics, which in our grandfathers day were relatively static, have now become fluid. (Cf. British Right and Left Wing approachment since Russia's entry into the war.)

   The clear picture of the political scene, which an agent should initially work out in his area, should not be like a chart, map or photograph with its details fixed and rigid, but like a moving film at a cinema where the frame remains fixed but contents are mobile.

b) Politics, to the propagandist, are not merely Party Politics; but a detailed examination of every factor affecting the self-interest of each group, class and organization that exists within his field of operation.

SECRET

6. GROUPINGS.

It is one of the early duties of a propagandist, even before leaving this country, to ascertain what groups exist in his area. This gives rise to two questions:

a) Why?

   Having ascertained a particular political fact that affects his area, the propagandist may easily find that each group of people has a different attitude to the same political fact.

   Therefore it follows that our propaganda relative to that one fact must differ vis-a-vis each group.

   E.g. Propaganda against agricultural plunder by the Boche will have to be distinct, if addressed to country folk who see the plunder happening, from that addressed to townsfolk who, perhaps, only know that their bellies are empty.

By addressing our propaganda to a relatively small specific group rather than to a large amorphous mass, we ensure that we shall appeal more directly to the self-interest of that group and therefore incite it the more easily to the action required.

This is, in fact, one of the main reasons why we send a propaganda agent to work inside a country since his leaflets, alone of all our propaganda <u>media</u>, can be addressed to and distributed among a specific group. B.B.C. broadcasts and R.A.F. leaflets cannot be thus addressed and distributed with certainty of success.

b. How?

SECRET

It is possible to group any given population under eight main headings:

Party Political. (Conservative, Liberal, Socialist, etc.)

Vocational. (Miner, lawyer, journalist, etc.)

Regional (Town, country, coast, etc.)

Religious. (Catholic, Protestant, Agnostic, etc.)

Age. (Students, youth movements, sports clubs, pension ex-servicemen, etc.)

Sex. (Workers wives, feminist movements, etc.)

Economic. (Employers, workers, trades unions, etc.)

National (Racial minorities.)

7. CHOICE OF TARGET.

SECRET

From the many groups at his disposal, our propagandist must choose the few groups that he can most profitably attack.

E.g. One might ignore extreme loyalists (already sufficiently active) and extreme pro-Nazis (already beyond conversion) in order to concentrate more effectively on the "attentiste" element of a population.

Once a specific group is chosen, the propagandist must determine (very broadly) which of the eight factors listed above is that particular group's predominant factor.

E.g. Socialists: predominantly a Party Political group.
      Students: predominantly an age group.

He must now ascertain whether the impact of the other seven factors on his chosen group is of sufficient importance to warrant splitting that group into smaller sub-groups.

E.g.   Q. "Does the fact that there are old, young and middle-aged socialists warrant my further splitting the main groups?"

        A. "Probably no."

        Q. "Does the fact that there are rich socialists (Theoreticians) and poor socialists (Practicians) warrant my

- 27 -

further splitting the main group?"

    A. "Probably yes."

8. CONCLUSION

With his population grouped and a specific group chosen as targets, the first preparational stage of our propagandist's work is complete.

Two further preparational steps remain to be taken:

a) The ascertainment of the chosen group's opinion on a given political fact or facts.

b) The receipt from a superior authority of a propaganda line or policy.

These two steps form the subjects of ensuing lectures.

D

CIPHER, Elementary Course
Prepared by GEOFFREY MOTT-Smith

LESSON #1

(NB - Material marked OPTIONAL is to be used only if (a) time permits and (b) interest and intelligence of the class warrants.)

1. This is a course in ciphers.

2. Do not at any time in this course take any notes. Close your notebooks and put them aside. Whatever you learn you must have up here (tap head) and anything that you do not have up here you must not use.

3. Scarcely anything is more incriminating, if found on the person of an undercover agent, than material of any description having to do with ciphers. If you are working under cover and must either send or receive messages in cipher, you must realize that every moment of the time you have them in your possession you are "HOT." As a matter of security it is to your interest to have a cipher message and the working sheets used to encipher it or decipher it on your person as short a time as possible. You must learn to work fast, secretly, and you must destroy by burning all these incriminating papers as soon as you can.

4. During this course we will make one exception and allow you to take down one written memorandum. It is essential that you know the English alphabet in what we call the NORMAL order -- A, B, C, etc. If you do not know that normal alphabet, copy it down and

learn it thoroughly.

5. I will write the alphabet on the blackboard, making the letters in the prescribed GI manner. I ask you all to copy it, making the letters in the same way. Write each letter wholly within a box of the paper, large enough to be easily readable.

(Write normal alphabet on board, using GI system, and pointing out features to be slightly exaggerated in order to avoid confusion of O, C, D, of U, V, etc.) (Supply one sheet cross-section paper to each man.)

6. I have no doubt that most of you know what a cipher is. A cipher is a written communication with no apparent meaning to the casual reader but which actually conveys a hidden message between two persons who know the key.

7. Before I continue I want to define a few terms so that you will understand what I am talking about. By "message", "Plain text", "cipher text" we mean the message you wish to convey to your correspondent. All three terms mean about the same thing. By "cipher text" we mean the jumble of letters that you actually send. (Write terms on board.)

8. By "encipherment" we mean the process of converting plain text into cipher text. (Draw arrow pointing from the one word to the other.) "Decipherment" is the reverse -- the conversion of cipher text to plain text (draw arrow.) Both of these terms imply that the encipherer or decipherer has full knowledge of the general system and the specific keys used. In other words, both of them are "in the know."

9. I will use one other term -- "Cryptanalysis" - which means the process of extracting the

- 30 -

intended message from a cipher without advance knowledge of its key. In all cipher work we must constantly remember the enemy cryptanalyst, who often obtains our cipher texts and tries to break them down.

10. (OPTIONAL) - Any written communication that may fall into the hands of the enemy MUST be in cipher. For your own security as under-cover agents I could wish you no better fate than that you never have to use cipher. But there are two common circumstances in which you cannot avoid it:

    FIRST - If you must use radio, as you usually do in sending reports to headquarters, all messages must be in cipher. The enemy of course listens in and picks up everything on the air. Here for your personal security you have to rely on concealing the source of your radio signals. Your radio operator as a rule has to find some lonely secluded spot, rap out his message, then run like hell before the detection finders spot him and the squad cars arrive. For the security of the whole organization, as well as yourself, the messages have to be enciphered in the hope that the system will defy the enemy cryptanalyst.

    SECOND - If you must communicate by mail, and have to say something incriminating, something to do with your undercover work, then your message must be enciphered and the cipher text must be written into a so-called "innocent text" letter which - we hope - conceals the fact that any cipher message is present.

11. No doubt many of you know that there are two types of operations that can be performed on a plain text to transform it into a cipher.

These two are SUBSTITUTION & TRANSPOSITION. (Write both words on board.)

12. Having noted that there is such a thing as SUBSTITUTION, I want you to forget all about it. We will not deal with any kind of substitution cipher. (Erase word from board.) We are concerned only with transposition ciphers.

13. (OPTIONAL) - In prescribing the cipher system you are to use, your headquarters can go only so far as regards SECURITY. The rest is up to you. The system we use, we believe, has a high degree of security from the enemy cryptanalyst, but you can destroy much of it by mistakes or by carelessness in using the system. On that account I want to show you precisely the elements of security in the system. To do that I will have to dip a little into cryptanalysis.

14. (OPTIONAL) - In English, as in every language, a few of the letters of the alphabet account for most of the words. The vowels A, E, I, O, and the consonants N, R, S, T make up about 60% of the letters in any random sample of English text, of say 200 or more letters. If you count the number of A's, the number of B's, etc. in a sample, you will find the result to be a kind of graph that looks like this (illustrate on board). About 13% of the total will be E's, about 10% T's, and then will follow AONIRS ranging from about 8% to 5%, with many possible variations of order. At the other end of the scale you will find rare letters: JKQZX, comprising all together about 1/2%. What we may call the "Graph" of the frequencies has a tilt that looks something like this. (draw line.)

15. (OPTIONAL) - What is the first thing the cryptanalyst does when he is given a cipher to break? He writes an alphabet A, B, C, etc. and counts the frequency of each letter in the cipher. Thus he compiles a graph of the frequencies, from which he gets his first clue as to the system of encipherment. Now, the ideal SUBSTITUTION cipher would give a perfectly flat graph (illustrate), and in fact a system other than the simplest gives a line much less tilted than this and with fewer blanks or no blanks here on the right. In fact, the only kind of text that gives exactly this very tilted line is plain-text itself. Your original plain-text message gives this kind of graph. If you encipher it by TRANSPOSITION, merely mixing up the order of the letters but not changing their identity, your cipher gives the same graph as the plain text. That graph can be recognized instantly. In short, when we use a transposition cipher, the enemy cryptanalyst knows that fact from the mere count of frequencies.

16. (OPTIONAL) - Furthermore, the enemy knows the history of cryptography as well as we do. He knows that while hundreds of transposition systems exist, only one type is practicable in military communication. Only one type has all the necessary qualifications - rapidity and ease in encipherment and decipherment, simplicity so that the system can be taught to thousands of cipher clerks, and a sufficient degree of security from cryptanalysis. This type is COLUMNAR TRANSPOSITION. The enemy knows that we use COLUMNAR TRANSPOSITION and we know that he uses columnar transposition. The sys-

tem was used in the last war and long before, and it will still be used after the war. (Write word COLUMNAR on board.)

17. To illustrate the system we use, COLUMNAR TRANSPOSITION, I will write a short message and make it into a cipher by a very elementary transposition.

```
        1 2 3 4 5
        T H I S I
        S A N E A
        S Y T R A
        N S P O S
        I T I O N
```

TSSNI HAYST INTPI SEROO IAASN

18. This is a message enciphered by COLUMNAR TRANSPOSITION. What we mean is simply that we wrote the message into a "encipherment box", wiring in the normal fashion from left to right and from top row down; then we copied out the columns, writing them horizontally in the usual fashion.

19. It should be evident to you that this is a very weak cipher indeed. The enemy cryptanalyst observes that there is a total of 25 letters. "Ha", says he, "perhaps the box was a 5 x 5 square." He tries filling the columns of the box in order, and he hits the message on the first try. What can we do to increase security, to delay his solution of the cipher? Well, one obvious thing we can do is to avoid using a square or a regular rectangle. We can write the message so that there is an "overhang" or "foot" down here (illustrate), so that some of

the columns are "short" and some are
"long". Then the enemy gets no clue from
the total length of the cipher.

20. That brings us to one of the basic rules
of our system:

21. NO ENCIPHERING BOX MAY BE A COMPLETE REC-
TANGLE - IT MUST ALWAYS HAVE A "FOOT".
The foot must be at least two columns
broad; at best it is about half the width
of the box.

21. Another evident improvement we can make
is to take the columns out, not in regular order from left to right, but in some
transposed order, for example

$$4\ 2\ 5\ 1\ 3$$

This numbering of columns which indicates
the order in which they are to be copied
out of the box is called the NUMERICAL
KEY.

22. We cannot conceal from the enemy that we
use columnar transposition. What is
secret from him? Well, I cannot empha-
size too much that our only secret is the
specific NUMERICAL KEYS, our main
reliance for security is upon frequent
change of keys, beyond that it is up to
you to safeguard the keys and the messages by avoiding certain errors which I
will point out to you.

23. (OPTIONAL) - It is an unfortunate fact
that a transposition cipher (much more
than a substitution cipher) is likely to
become wholly indecipherable if an error
is made during encipherment. In fact,
GARBLES that take much time to unravel
are so frequent when the system is used

by inexperienced or careless operators that we have to spend considerable time in this course on the study of UNGARBLING.

24. (OPTIONAL) - You yourself may be an entirely accurate encipherer. I hope you will be by the time you finish this course. Yet you may have to deal with sub-agents to whom you may have to teach a cipher system, and you must be prepared to deal with the garbles they are likely to make.

25. Throughout this course there are two watchwords to be borne in mind - ACCURACY & SECURITY. You must be accurate in the mechanical operations of encipherment and decipherment, to avoid time-wasting garbles. To protect yourself and the system you must faithfully observe all prescribed cautions in constructing your messages, and you must also guard against being found with incriminating cipher material on your person or among your chattels.

26. To drive home to you how important it is to keep track of your working sheets and be sure that all are finally destroyed, I am going to parcel out this cross-section paper one sheet at a time, and all sheets must eventually be returned to me for destruction. There will be a supply of paper available to you for your night problems. On taking any of this paper you must sign attached slip and note the number of sheets you take.

27. You will also receive some mimeographed sheets of problems. These sheets are secret and all must eventually be returned to me. I hope they will be in good con-

dition. No cipher material of any description may be out of your immediate possession except that it may be left in the box in your room. Do not leave working sheets in your notebooks and then leave your notebooks in the lounge. And of course do not take any of the material out of this area. It is all secret matter. Do not keep any notes of your own about the cipher system, excepting only that you may have a written memorandum of the normal English alphabet.

28. Now let's return to the subject of NUMERICAL KEYS. It is manifestly impractical to ask you to remember a disordered sequence of numbers, especially when the sequence may run up to twenty or thirty figures and when it is continually changed. It is easy, however, to remember a word or phrase. All our numerical keys are therefore derived from a series of words. The rule is: Number the letters of the key from 1 up, taking the letters in the order in which they come in the NORMAL English alphabet. To illustrate:

```
W   A   S   H   I   N   G   T   O   N
10  1   8   3   4   5   2   9   7   6
```

Here we have two N's. When the same letter occurs two or more times, we number these occurrences consecutively, beginning at the left. (More examples if necessary.)

29. (OPTIONAL) - If you do not know the normal alphabet well, or if you find you have trouble deriving numerical keys, first write out the alphabet, then tackle the key-phrase.

30.   In the mechanical operations of encipher-
        ment and decipherment I will ask you to
        follow a routine and to make certain
        checks as you go along. You will dis-
        cover the value of this routine the
        first time you make a serious error. In
        deriving a numerical key, circle the last
        number you write down (illustrate.) This
        number should check with the number of
        letters in the key-phrase; count the let-
        ters to see that it does. Place verti-
        cal marks through the two rows dividing
        the key-letters into groups of five
        (illustrate). These marks in fact ought
        to be put in before you commence to
        derive the numerical key. You will see
        that one utility of grouping the key-let-
        ters in fives is to give you a quick
        count of the total which you will check
        against the last number - the highest
        number - in your numerical key.
        (Illustrate).

31.   Now I am going to give you eight different
        key-phrases and I will ask you to derive
        the numerical key from each. Copy the
        phrases on separate rows, leaving at least
        two blank rows between each pair - better
        make it three or four blank rows. Before
        you start, let me caution you to try for
        speed, but not at the expense of accuracy.
        If you find you have made a mistake, and
        the mistake involves several numbers,
        cross out the whole row and start over
        again on the next blank row below. It is
        usually a waste of time to try to correct
        by erasing if you can cross out and start
        over.

        H O R S E F E A T H E R S
        6 8 9 11 2 5 3 1 13 7 4 10 12

```
F  U  R  L  O  U  G  H  G  R  A  N  T  E  D                SECRET
4 14 11  8 10 15  5  7  6 12  1  9 13  3  2

K  A  T  Z  E  N  J  A  M  M  E  R  K  I  D  S
8  1 15 16  4 12  7  2 10 11  5 13  9  6  3 14

D  O  N  T  G  I  V  E  U  P  T  H  E  S  H  I  P
1 10  9 14  4  7 17  2 16 11 15  5  3 13  6  8 12

J  U  M  P  Q  U  I  C  K  B  R  O  W  N  F  O  X
5 14  7 11 12 15  4  2  6  1 13  9 16  8  3 10 17

I  A  M  N  O  T  A  T  E  N  P  E  R  C  E  N  T  M  A  N
8  1  9 11 15 18  2 19  5 12 16  6 17  4  7 13 20 10  3 14

E  I  G  H  T  H  U  N  D  R  E  D  T  W  E  N  T  Y  S  I  X
3  9  6  7 15  8 18 11  1 13  4  2 16 19  5 12 17 21 14 10 20

Q  U  I  C  K  S  E  R  V  I  C  E  G  U  A  R  A  N  T  E  E  D
15 20 11  3 13 18  6 16 22 12  4  7 10 21  1 17  2 14 19  8  9  5
```

(Allow time for first one or two men to finish. Stop all men at same time, require papers to be turned over. Thus get indication of who is fast and who is slow.)

(Break here if necessary, but try to keep a tempo that will all break to be deferred to next point indicated.)

32. Now let's go through an example encipherment. I will ask you to copy the example as I write it on the board. For our first key we will take MARYLAND. Write that near the upper left corner of your paper and derive the numerical key.

33. Draw vertical lines enclosing the key. These lines mark the width of the encipherment box. We can't mark the bottom of the box until we know the length of our message. Let's take the message:

WRITE MESSAGE IN BY ROWS COPY OUT BY COLUMNS

We write it in the box in normal fashion, from left to right and from top row down. Let me caution you to leave no blank spaces. Fill every cell as you go along. Don't make any break between words.

```
M A R Y L A N D
5 1 7 8 4 2 6 3

W R I T E M E S
S A G E I N B Y
R O W S C O P Y
O U T B Y C O L
  U M N S
```

SECRET

34. For the sake of radio transmission, as well as other purposes, we have a rule:

    THE TOTAL OF LETTERS IN THE MESSAGE MUST BE MULTIPLE OF 5. IF NECESSARY, IT MUST BE FILLED OUT TO A MULTIPLE OF 5 BY THE ADDITION OF NULLS AT THE END.

    By nulls we mean letters that have no meaning so far as the message is concerned, but are added merely to lengthen or break up the message.

35. Just a word on the choice of nulls. The enemy cryptanalyst is greatly helped if he can spot the letter or letters that either begin or end the message. We therefore avoid adding as nulls at either end those letters which LOOK like nulls. We avoid the rare letters. The safe rule is to choose nulls mostly from "the Big Eight" letters - AEIO and NRST. But don't fall into habits in your choice of

nulls. Don't adopt a stereotyped formula. Mix up your practice. Use middle-frequency letters occasionally, such as B, V, G, F, P etc. Simply see that "Big Eight" letters predominate. If you have to put in a great many nulls, say more than six, put in a word that simply has no connection with the sense of the message, or even add a phrase such as YANKS WIN BALL GAME.

36. Our message contains 36 letters. If we add just four nulls to bring the total to 40, we shall violate the first rule by making a complete rectangle. Therefore we must either cut one letter out to make a total of 35 or add 9 nulls to provide a "foot" on the box. Let's add 9 nulls, and let's all use the same series as follows: AEIONRSTE.

37. (OPTIONAL) - At this point we could make a simple columnar transposition by copying out the columns in order and using the result for our cipher text. I am going to do just that on the blackboard but I don't want you to copy it down. Just watch for a while.

```
        M   A   R   Y   L   A   N   D
        5   1   7   8   4   2   6   3
        _____
```
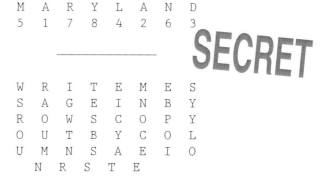
```
        W   R   I   T   E   M   E   S
        S   A   G   E   I   N   B   Y
        R   O   W   S   C   O   P   Y
        O   U   T   B   Y   C   O   L
        U   M   N   S   A   E   I   O
            N   R   S   T   E
```

RAOUM RMNOC ESYYL OEICY AEWSR OUNES POIIG WTNST ESBST

38.   (OPTIONAL) - Here is a cipher produced by a single columnar transposition. I want you to understand that this is one of the easiest types of cipher to break. It is important for you to realize that fact. The system we use commences with just this sort of transposition and once in a while a hasty operator forgets to go through with the rest of it. He sends a single transposition cipher and so gives the enemy cryptanalyst a text that he can read within a few minutes.

39.   (OPTIONAL) - The cryptanalyst does not know the size of the enciphering box, the length of the columns, but he does know that if this cipher is a single columnar transposition its first letter was the top of some column, and a certain number of letters commencing from the head of the cipher formed a complete column. Likewise the end of the cipher marks the foot of some column. Now, somewhere inside the cipher are the columns that stood next to these columns that were copied out first or last. The analyst picks these out in a matter of minutes, perhaps seconds.

40.   (OPTIONAL) - What he does is to copy off six to a dozen letters at the beginning or end on a separate strip, then he slides this strip along against the rest of the cipher until he finds the place where it fits best. (Illustrate.) What he is looking for is good English digraphs, combinations of two letters. Any two adjacent columns of your encipherment box will show a reasonable number of good English digraphs, such as TH, ER, AN, etc. This slip cannot fit well in more than a few places, and usu-

ally, if the columns are not too short, it fits well only at one place - the patch which is the column that stood next to it in the box.

41. (OPTIONAL) - Having found a patch that fits well with the patch on the strip, the analyst can write the two together on a strip and match them with the rest of the cipher to find the next adjacent column. He is on very sure ground because the digraphs usually fix some of the letters that must appear in the adjacent columns. Thus by matching the analyst can break down the cipher into its original columns. But more than that. When he finds the very first pair of columns, he has strong indication of the dimensions of the encipherment box. There will be only a few sizes of box that will let this letter (indicate) which matches with the first letter of the cipher be also the head of a column.

42. A single columnar transposition is actually about the easiest type of cipher to break. We would not dream of copying out RAOUM, etc., and sending that as the cipher. Be careful that you do not send that sort of thing by accident.

43. What we actually do is to copy out RAOUM, etc., and put it through a SECOND transposition. Our system is a TWO-PHASE or DOUBLE transposition.

SECRET

44. Let's take for our second key ARIZONA. Write that in the upper right of your paper and derive the numerical key.

```
A R I Z O N A
1 6 3 7 5 4 2
```

45. Now let's go back to the first box. Count the length of column on the right, the short column. It is 5. Write 5 beside the right-hand margin (illustrate). Count the number of letters in the foot. Here it is 5. Write 5 just below the foot. Don't omit these figures, even though you can make all calculation in your head. You will find those check figures mighty handy the first time you make a serious error under pressure (expand ad lib.)

46. This figure 5 (right margin) times the circled figure 8 (in key) plus the foot 5, or 45, gives you the number of letters in the message. Always check that total after you finish the first box.

47. The second box is going to be 7 columns wide. On the side somewhere make this division. Write it all out, even though you can do it in your head.

```
  7) 45 (6
     42
     ──
      3
```

SECRET

48. The dividend 6 indicates that your second box must be 6 deep on the right, the short column must be six rows long. Draw the right margin line down to this length. On the left, the margin line will go one row deeper, and the remainder 3 indicates that you must have three spaces in the foot (draw second box.) That gives you the exact dimension of the second box.

49. Working in a hurry in the field, you may sometimes get confused as to how to write in and how to take out. Here is a rule to remember to keep you straight; whether you are ENciphering or DEciphering, write letters into both boxes in the same way, and copy out of both boxes in the same way. Now, you remember that we always write the message into the first box in the normal way - from left to right and from the top row down. Then, write the letters into the second box in the same way. To get a transposition, take out of the first box by columns. That is where you numerical key comes in. Copy the columns out in accordance with the key, column 1 first, column 2 next, etc. I ask you to copy the letters into your second box and then check with what I will write on the board.

50. Here is another check you should not fail to make. In copying from the first box to the second, UNDERLINE (illustrate) each letter which is the bottom of a column. When you come to the end of the second box, if you have a letter too many or too few, go back and check the columns by counting from one underline to the next (illustrate).

51. We are now ready to write the cipher text. Copy out the columns of box two in accordance with its numerical key. Write the letters in rows, as you would in writing a letter. But group the letters in fives. You can do that either by leaving a space between each group or by first putting vertical marks to divide the sheet into blocks of five columns each (illustrate). SECRET

52. (OPTIONAL) - With a very long message, it may be advisable to run down the columns of box two in order and circle every fifth letter. Then when you copy out, make breaks after each circle. Thus you guard against having to do a hell of a lot of recopying by reason of an early mistake in grouping.

53. There are 45 letters in the message. After copying out the cipher text, count to see that you have 9 groups of 5.

54. To every cipher message we must attach a group of 5 numerals, which we set at the head and repeat again at the end. In this group the first two digits give the number of the message. You will number your message from 1 to 100. As you must always use exactly two digits, the numbers 1 to nine must have a zero prefixed, thus 01, 02, etc. After you reach 100, you start over again at 01.
(OPTIONAL) I may say that the number of your message is in fact an indication to your correspondent of the two keys you have used. The system provides you with a way of remembering 100 keys, which you use in regular order.

55. The last three digits of the numeral group gives the number of letters in the message. In this case we have 45 letters, and as we must use three digits we must write it as 045. We will suppose that this is our first cipher message, so that our numeral group is 01045. Write that group at the head and at the end of the cipher text and the message is ready to hand over to your radio operator.

SECRET

(BREAK) **SECRET**

56. Now we will run through the process of decipherment. Obviously this process is simply the reverse of encipherment. But I want you to learn certain points in the routine.

57. You have received a cipher message, such as this (encipherment made earlier.) What is the first thing to do? Well, look at the numeral group, count the cipher groups, and see that they agree as to the number of letters. In radio transmission it is not uncommon for a group to be dropped or missed, and if you are going to have to deal with an incomplete text you want to know it at once.

58. The text is all here. Next you look at the first two numerals and they tell you the keys. Write the two keys, left and right on your sheet, and derive your numerical keys. (NOTE: At the outset, have box one placed on left, box two on right, as in encipherment. Sooner or later a student will complain that then in copying from two to one he finds that his right hand gets in the way. Take this opportunity to state that this layout, AS WELL AS OTHER POINTS OF THE ROUTINE, are recommended for clarity, but that the student may and should make modifications to suit himself whenever he sees a way to speed up his own work.

59. Before entering a single letter in the box, draw up both boxes. Write out the divisions in full.

        (For box one)   (For box two)

```
            8 )  45   ( 5        7 )  45   ( 6
                 40                    42
                  5                     3
```
SECRET

(Make boxes.)

60. Remember that the last step in encipherment was to copy out the columns of box two. The first step in decipherment is therefore to write the cipher into the columns of box two. I will ask you to do that, then check with what I shall write on the board.

61. The rule is that we write into both boxes in the same way. We wrote into box two by columns. Therefore we must copy into box one by columns, taking out of box two by rows in the normal fashion of reading. Please do that now.

62. When you write in box one, underline the letters that come at the end of each row in box two. (OPTIONAL) Let's go back a moment to the first step. When you write the cipher into box two, underline each letter that marks the end of a group.

63. With the letters copied from box two into box one, you can read the plain text on the rows of box one. (OPTIONAL) In the field, it is usually sufficient for you to read the message out of the box making word-breaks by eye. You will usually want to burn your papers as soon as you have grasped the content of the message. In this course, however, I will ask you to indicate that you have understood the message, either by marking the word-breaks in box one with heavy lines or by copying out the plain text.

(Collect all papers)

64. We will now take up some questions of cryptographic security. Look at the underlines in the second encipherment box, which mark the ends of the columns of the first box. These underlines are useful in tracking down errors. But they are also important in enabling you to estimate the degree of security achieved by the second transposition. Remember that the first transposition makes a cipher very easy to break. The second transposition makes a cipher very hard provided that it is carefully done.

65. For the most effective second transposition, the underlines in box two should be well scattered, with no concentration in any one column. (OPTIONAL - Ad lib on the reason why.) If you see any considerable number of underlines on one column, and especially if more than half are in one column, you should tear up your work and start over again.

66. One way in which this concentration of underlines in one column comes about is that some columns of box one - either the long or short - are the same length as the rows of box two. Your box one columns are copied off into the rows of box two, and the last thing you want to have happen is for the column to fit exactly into the row. It should break before the end of the row or overrun it into the next row. So we come to another law of the Medes and the Persians:

    NEVER LET A COLUMN OF BOX ONE BE THE SAME LENGTH AS THE WIDTH OF BOX TWO.

67. The best way to avoid breaking the rule is this. Set up your box one. You know

your keys, you know the width of box two. Suppose the width is 7. Count down seven rows below the first key. On the 7th row, you must now end the message. To be on the safe side cut back two rows - below the 5th row draw a heavy line. Also go forward two rows and below the 9th row draw a heavy line. The area from the 5th to the 9th is FORBIDDEN - ye shall not end the message in this area.

68. If you find that what you have to say is going to end up in the danger zone, add enough nulls to carry you out of it. That may involve a great many nulls. Then it is simpler, instead of stringing out disconnected letters, to write some nonsensical words of a phrase that has no connection with the real message. Say HUMPTY DUMPTY SAT ON A WALL or whatever you will.

69. Now some points on writing the message. In the first place, we must note that one way of tackling the cryptanalysis of a transposition cipher is simple anagramming. Suppose I tell you that this assortment of letters can be made into one word

                NNHSGTWOIA   **SECRET**

What is the word? (Let class answer) WASHINGTON - yes. You got the answer by trying out various combinations of letters until they suggested a word. Well, the cryptanalyst can try exactly the same process, and with a short cipher he may succeed. It may seem incredible that anagramming can work with 50 or 60 letters, but remember that many common

English digraphs and trigraphs can be expected to appear in a message of this length. By taking out such combinations as TH, ER, ENT, & TION, the analyst may find that the residue of letters suggest certain words and thereby certain phrases.

70. To defeat simple anagramming we have a rule

    DO NOT WRITE LESS THAN 100 LETTERS IN THE MESSAGE

71. We also find that a top limit is necessary. You may have a radio operator to consider. As a matter of security you may have to give him a message short enough so that he can rap it out and get away in time to escape the detection finders. Then, too, the longer the message, the greater the danger from a garble. So we say

    AS A RULE, DO NOT WRITE MORE THAN 300 LETTERS IN ONE MESSAGE

    If you have much more to say, break it into two or more messages. One message then can vary within the limits of 100 to 300 letters.

72. The enemy cryptanalyst is helped greatly if he can spot the letter that begins or that ends the plain-text message. He tries to find these points by trying out probable words. We try to defeat him by avoiding stereotyped formulas at either end. Don't start out with a salutation; don't end up with a signature. If you must include some common formula such as "John to Peter" or "In reference to your number seventy-three", shove this phrase in the middle of

the message somewhere - anywhere - even in the middle of a word. To make clear that it is parenthetical, or that it should be taken out and applied to the message as a whole, set it off by the letter X at both ends. (Illustrate)

73. The letter X is an all purpose letter, used for any kind of break mark of punctuation, etc. Particularly avoid spelling out STOP, COMMA, and so on. Use X instead. From the cryptanalytic point of view, the X is no weakness. Since we use it for so many purposes, the enemy cryptanalyst doesn't know where any particular X comes into the message and gets no information from it.

74. But he may get entrance if we have too many letters that are "choosey" in their contacts. All letters have preferences as to the letters that precede or follow them, but some are much more particular than others. The choosiest is of course Q, which is always followed by U. Therefore we never use Q, and we also dispense with a couple of other rare letters which are not really necessary.

FOR Q WRITE C: FOR J WRITE I: FOR Z WRITE S

(OPTIONAL) - Remember that in early days J & I were the same letter anyhow. We don't like Z because its only frequent use in military cipher is in the word ZERO. Therefore we write SERO.

75. Do not put numerals in the message. All numbers must be spelled out, for example SIX THREE SERO.

76. Use abbreviations if they will be under-

stood by your correspondent. For example, HQ is generally understood for HEADQUARTERS. But we have to write it HC under the rule. If the preceding word ends in T, there might be doubt as to where the word-breaks come. I take this simply as one example of possible ambiguity. To overcome such ambiguity, or to make sure that we get across some word or name whose exact spelling is important we may use the following device. The group RPT means REPEAT. We put RPT after the crucial word and immediately repeat it, as HC RPT HC. This shows the first and last letters of a group separated out of the context by word breaks, or confirms a doubtful spelling.

77. You will notice that we have not prohibited the use of the term letter K. It is still a good idea to omit K where there is no doubt as to the word intended. For example ATTAC will usually be understood clearly as ATTACK. Similarly TRUCS is clear. By the same sign, don't worry about minor misspelling. If you have trouble with such words as RECEIVE, don't bother about it. A little judicious misspelling makes it tougher for the enemy cryptanalyst. Such a word as ACCOMMODATE, as the intention is clear. I have given this advice to previous classes, and have found that some students have taken it for license to write Egyptian hieroglyphics. Remember at all times that your meaning must be clear (CLEAR) to your correspondent. Don't make him guess what you mean. Especially if you are dealing with a sub-agent, some man who has not had thorough training or much experience in ciphers, you had better make your language simple and avoid

all cryptographic tricks. (peroration ad lib.) (Assign Problems 1 & 2 for night work. Place a pile of cross-section paper in an available place, with a separate sheet which each student must sign and state number of sheets taken.)

SYNOPSIS of course on CIPHERS
in four Lessons, for E area

Prepared by Geoffrey Mott-Smith

FIRST LESSON **SECRET**

1. Define:
   Cipher   plain text   cipher text
   message   clear text   crytanalysis
   encipherment   decipherment   cryptography

2. Basic types of encipherment operations are (a) Substitution (b) Transposition (Now forget all about Substitution. System you will learn is Transposition.)

3. Watchwords to be borne in mind throughout are

   ACCURACY - messages must be decipherable by intended recipient (Transp. method is unfortunately more affected by garbles than Substitution.)
   SECURITY - strict adherence to all cautions in use of system, for self-preservation and also safety of whole organization.

4. Basis of security of our system cryptanalysis:
   (a) Any transp. cipher is recognizable at sight;

- (b) Security therefore rests entirely upon secrecy of specific keys;
- (c) Enemy knows that only transp. system practicable in military communications is COLUMNAR TRANSP. (Used by Germany at outset of last war; in widespread use today.)
- (d) Careless cryptography can facilitate task of cryptanalysis.

5. Example of simple columnar transp. (THIS IS AN EASY TRANSPOSITION in 5x5; columns copied out in normal order.)

6. Obvious ways to increase security of this type:
   - (a) Always have a "foot" on the encipherment box;
   - (b) Take out columns by a mixed numerical key.

7. Use of words and phrases for keys:
   - (a) Easy system is arranged for memorization of 100 keys at a time (not to be taught in this course.)
   - (b) Derivation of numerical key from words.

   CLASS EXERCISE - Derive keys from 8 phrases of differing length and character.

8. Example encipherment, simple columnar transp. (Perform on blackboard, have class copy on paper.)
   THIS TYPE OF CIPHER IS ONE OF EASIEST TO BREAK. (Optional: give demonstration of matching of columns.)

9. System we use is TWO-PHRASE or DOUBLE transp. Continue example by a second transp. (Class copies from board.) (Optional: why the cipher now becomes much more difficult to break.)

10. Arrangement of cipher text:
   (a) Grouping into pentagraphs;
   (b) Numerical group written at beginning and end.

11. Basic rules for security in encipherment:
   (a) Set up first box, mark out "danger zone" where message must not end (centered by width of second box)
   (b) After writing message, add nulls to make total a multiple of 5;
   (c) Choose nulls predominantly from high-frequency letters;
   (d) Each box must always have a foot, which at best is about half the width of the box - never less than two columns;
   (e) Length 100 - 300.

SECRET

12. Mechanical checks on the layout:
   (a) To determine size of box (total elements being fixed) write out long division (dividend is length of short col; divisor is width of box; remainder is width of foot.)
   (b) In copying from Box 1 to Box 2, underline last letters at bottom of Box 1 cols.
   (c) Take out of Box 2 by pentagraphs (if message is long, mark out pentagraphs in Box 2 before copying out.

13. Cautions in message writing:
   (a) Instead of J, Q, Z, use respectively I, C, S.
   (b) Use X as all-purpose letter, for punctuation, etc.
   (c) Avoid stereotyped phrases at beginning or end; put in middle any necessary formulas, set off by X's.
   (d) Optional Use of RPT symbol; use of abbreviations.

> (e) Misspelling helps to defeat crypt-
> analysis, but be sure that message
> is clear to recipient. (Don't play
> tricks in writing to inexperienced
> agent.)

14. Decipherment (If time permits, give short
cipher for class to decipher; otherwise,
show on board decipherment of message
previously enciphered.)
    - (a) Lay out both boxes first, write out
    long divisions; write auxiliary
    numerals;
    - (b) Reverse steps of encipherment;
    - (c) In copying from 2 Box to Box 1,
    underline end letters of Box 2
    rows.

NIGHTWORK: Problem 1: decipher given text.
Problem 2: encipher given text.

SECOND LESSON

Message-writing and encipherment test. Have class write a cipher message, conveying certain specified information. Each man may choose his own keys. (Optional: write on blackboard some suggested keys.) Caution that keys must be at least 10 long.

Message need not respect rule of 100 minimum. Urge all possible brevity.

Collect encipherments, redistribute so that each man has work of another, and require decipherment. (Cipher texts must be written out on separate sheets together with notation of word-keys.)

Introduction to garbles. What they are, some ways they arise. Degree of garble, and often its nature, recognizable from amount of plain text in Box 1 after decipherment. Use common sense.

NIGHTWORK: Problem 4, decipher given text, garbled by misnumbering two consecutive columns of Box 1
(Optional: extra problem, encipher given text.)

If time and intelligence of class permits, in this lesson explain shortcut method of deciphering when one pentagraph has been lost from cipher text, and assign.

> Problem 5: decipher given text from which one pentagraph is lost.

THIRD LESSON

Encipherment-decipherment test as in second lesson. But larger list of items is to be included in message, which must comprise at least 150 letters. Repeat all cautions as to message-writing and security in encipherment. (Usually it will be necessary to review at some length the errors made in Second Lesson.)

Review of methods of attacking simple garbles as applied to nightwork handed in.

NIGHTWORK: (NB: Schedule at E usually gives no time for nightwork after this lesson. If time is available, assign Problem 5 if not previously assigned, or else Problems 6, 7 encipherment or decipherment ad lib according to what the class needs to practice.)

SECRET

FOURTH LESSON

"INNOCENT TEXT" Letter - explain system. Caution that this may have to be explained by agent himself to a subagent, and that he may modify details to suit occasion.

- 58 -

Read example letter. Then require class to write such a letter, using "cipher text" ABCDE-FGHILMNOPRSTUVWY.

Read all letters aloud to class, require vote on each one as to whether it sounds as though it will pass censor. (Letters-writing cannot be taught; if a letter is here voted suspicious, writer had better not attempt the system at all.)

NIGHTWORK: problems ad lib according to needs of class. (May include Problem 9, decipher given text in which two consecutive columns of Box 2 were misnumbered. Warn of this fact. If time permits, give brief blackboard demonstration of how to trace back from Box 1 to Box 2 to discover the error.)

SECRET

John Gerry.
S&T, Area E.

E

SECRET

UNDERGROUND IN THE FAR EAST

(Oppositional trends - underground - guerrilla)
(This survey completes the lecture which develops the basic principles of Underground. It gives only a few items and data, mainly for the purpose of pointing out <u>in principle</u> the possibilities existing in that theater, possibilities which may be exploited by our organizers for SI and MO operations.
Naturally, this general has to be followed by detailed area studies.)

<u>KOREA</u>

The Japanese have not succeeded to crush the oppositional movement in Korea which is supported by a government-in-exile. This government, established in Chungking, is the oldest of all governments-in-exile: the KOREAN PROVISIONAL REPUBLIC, as it is called, exists since about 25 years.

The independence of Korea has been recognized in principle by the Conference in Cairo between Churchill and Chiang Kai-chek.

The Korean Provisional Government has at its disposal an Independent Korean Army of 10,000 men, stationed in Free China.

Some facts of the resistance in Korea:

Within 24 hours of the Doolittle Raid on Tokyo, a band of 14 Koreans lost their lives blowing up Japanese oil tanks near the Chemulpo Bay. (It may be expected that, with further successes of our Armed Forces against the Japanese, the resistance of the Korean population may be encouraged and take new and more efficient forms.)

In 1937, Japanese authorities listed 3600 instances of sabotage and guerrilla activities in Korea.

In some regions exists a passive resistance of farmers who are withholding agricultural products, and a non-cooperation in adapting agricultural production to Japanese needs.

100,000 guerrillas have been reported to live in the mountains of Korea.

Altogether there seem to exist good possibilities for an underground movement in this country, which has 22 million inhabitants.

In addition, it should not be forgotten that 2 million Koreans live outside Korea proper, in China and Manchuria. Among them we may find potential informers and sub-agents for those areas and also people who could work in Korea and penetrate Japan proper.

The greatest care will have to be used, of course, since the Japanese use often Koreans as informers and spies, also as Volunteer Auxiliaries for the KEMPEI (Gendarmerie). It is well possible that, once we are engaged in operations in China, the Japanese will try to penetrate our network in using Koreans as agent provocateurs.

P.S. The facts mentioned above, seem to be confirmed to a certain extent by a recent radio broadcast from Tokyo which admitted that the Government has difficulties with the Korean population.

SECRET

CHINA (FREE CHINA)

The activities of Chinese guerrillas in occupied China are known. Fighting now since seven years, they hold usually the territory between lines of communication held by the Japanese.

No given point in occupied China is more than 10 or 15 miles from guerrilla lines. The guerrillas are known to enter and leave most occupied towns at will and have lines not further than 20 miles from the very heart of Shanghai. (However, it should not be forgotten that there are also free-lance guerrillas who play both sides and depend on looting for a livelihood. Furthermore the Nanking puppet government has organized counter-guerrillas who have successfully impersonated Chunking men.

Chinese are also active in the Philippines where large numbers of them have been sentenced to death for espionage against the Japanese.

PHILIPPINES

Nothing is yet reported of new oppositional trends. But the guerrillas, fighting now since 1942, could not be crushed by the Japanese, though many have been caught.

MALAYA

No report of any oppositional trends or

activities. Obviously the anti-British attitude of the natives is too strong to be upset by the treatment which the Japs give the natives. Though discontent with Jap methods has been reported, it has not found at present any form of an organized opposition.

But Chinese guerrillas, led by British officers, are reported to operate now against the Japs in Malaya.

SECRET

BURMA

No sign yet of an organized opposition against the Japanese. Our operational units however could find the support of Burmese natives.

FRENCH INDO-CHINA

There exists a Free French underground movement. It will be difficult to overcome the anti-white attitude of the large majority of the natives, in view of the often criticized colonial methods of the French.

Recent reports, however, speak already of the formation of a French-Native anti-Japanese block. No results yet can be observed.

Another report mentions that many natives have recently crossed the border to Kwangsi where they are trained to fight against the Japs.

THAILAND

There seem to exist good possibilities for underground work and for support of our operations by the opposition. The large majority of the Thai people disapprove of collaboration and is in favor of the Free Thai Movement organized abroad, with headquarters in Washington.

After the occupation of Thailand by Japan, several independent underground groups arose in Thailand to organize sabotage against Japanese communications, guerrilla activities in the jungle, etc. Since 1941 train wrecks have not stopped.

SECRET

DUTCH EAST INDIES

The guerrilla groups of fighting Dutchmen which opposed the Japanese invaders since the occupation of the islands, have been crushed in several islands. Only in South Borneo seems still a resistance center to exist.

However, there are indications (in the Jap radio) of new anti-Japanese propaganda among the natives.

MANCHURIA

Manchurian, Chinese and Korean guerrillas have been fighting since many years in Manchuria. However, since then those guerrillas seem to be much weaker than they were before. But as late as April 1944, the Manchukuo Government admitted the existence of sabotage - and guerrilla - groups.

JAPAN

We cannot speak at present of an underground movement in Japan. Formerly such movements existed, in particular among liberal intellectuals and laborers. At the time of the

Manchuria "Incident" existed an anti-imperialistic league organized among students and industrial workers.

The elections of 1937 expressed the popular distrust of militaristic policy. (The Social Mass Party then won about 37 seats).

There were still major strikes 2 and 3 years ago with participation of 20,000 workers in one single enterprise.

A strong police and counter-espionage system, assisted by the "thought control" of the fascist movement, have succeeded to crush the formerly existing movements, and try to prevent the organization of new groups.

But there are signs of at least oppositional trends which, as events pass by, may be developed to real organized opposition and underground movements:

Again and again we hear of arrests of liberal professors and students.

Prof. Sakimura, a Japanese diplomat in Berlin, flew to Sweden and placed himself under the protection of the Swedish Government.

A Gripsholm repatriate reported to our G-2 dissatisfaction among laborers. As illustration was reported that, when Prime Minister Tojo went to visit a munitions plant, he barely escaped injury when a hammer was thrown at him.

There seems also to exist a nucleus of resistance of some Christian elements, though the official Christian Churches are cooperating with the regime.

When our troops occupied Kiska Island,

they found Japanese underground leaflets distributed among Japanese soldiers and attacking Japanese imperialism.

In Chungking and in Japanese-controlled North China the Chinese organized groups of anti-imperialistic Japanese prisoners-of-war who sent through underground channels letters and pamphlets to Japan.

Thus there seem to be chances that those oppositional trends may be developed to an important movement.

In this respect the Chinese seem to be hopeful, or at least they try to develop those trends.

Chiang Kai-chek invited recently the Japanese to revolt and stated the agreement between himself and the President of the United States: if the Japanese should revolt and overthrow the militarist government, their choice of a form of government should be free.

F

PROPAGANDA PRESENTATION

1. Fundamental Principles

INTRODUCTION

    Good advertising is based on set principles; good writing on deep feeling. Good propaganda needs both; and the good propagandist will use the latter to mask the former.

    The principles must never be allowed to "show through". We are not sending men back to Occupied Europe to sell soap.

    Nevertheless propaganda presentation must always conform to the following fundamental principles:

1. <u>Simplicity of</u>:

    a) <u>General Idea.</u>

        Let each leaflet present but one General Idea - from which there should be no digression.

    b) <u>Argument.</u>

        In support of the General Idea one may produce Particular Ideas. These should be logically linked and linked so closely that the reader is unable to escape from climbing the rigid "mental stairway" that leads from an existing attitude to a required attitude.

Such linking is most effective when it binds sentence to sentence (for examples, See Appendix A): but, where this is impossible, one should always link Particular Idea to Particular Idea.

    c) <u>Language.</u>

Be simple, but never patronising. Do not speak as a scholar writing down to fishermen. Lower your mentality to that of a fisherman and write up. (For example, see Appendix B.)

2. <u>Concreteness.</u>

    a) Avoid abstract words like the plague, because:

        1. Such words as "democracy", "patriotism", "freedom", have become platitudes without significance.

        2. Even where abstracts are not yet platitudinous, they can never affect a reader's self-interest so powerfully as concrete words. E.g.:

        For "patriotism" say "Love of France."

        For "hunger" say "empty bellies."

        For "The Peace Loving Dutch nation are now resisting German oppression" say "The Dutch people, who once grew tulips and made cheese, are now stabbing Germans in the back."

> For "Germany's death-rate is rising in Russia" say "German corpse is piled upon German corpse among the blood, the bone, the twisted tripes and scattered bowels of the Russian battlefield."

b) In thus appealing to a reader's self-interest, recall that such self-interest is normally two-fold:

1. Selfish: "The Germans have taken away your cattle."

2. Unselfish: "The Germans have enslaved France."

The most powerful possible appeal is one directed towards a combination of selfish and unselfish elements, i.e., "Your cattle have been taken away because France is enslaved."

3. <u>Repetition.</u>

a) By this is meant the choice of one urgent, compelling General Idea, and the repetition of that idea from many different points of view and/or by many different methods - leaflets, broadcast, rumor, etc.

b) By constant repetition we can be certain of securing:

1. A larger audience.

2. If not a conviction, at least an effect on the mentality of the audience. Cf. Goebbels anti-Semitic and anti-Czech propaganda.

4. <u>Action.</u>

Action must always be recommended because:

a) Action is the aim of all propaganda.

b) Action drives home the persuasive lessons of repetitive propaganda - i.e. people will remember a propagandist's line of talk if they can associate it with an action in which they themselves have taken part. E.g.

   1. Germans will remember Goebbels anti-Semitic propaganda because they have seen, read of or taken part in Jewish persecutions.

   2. They will remember his anti-Czech propaganda because Germany won a "bloodless" victory over the Czechs.

   3. They do NOT remember Goebbels anti-Italian propaganda in July 1934, because nothing was <u>done</u> about it.

G

OPINION SAMPLING

NOTE: This lecture deals with an activity which can only be engaged in on definite orders and in relation to the specific security arrangements made in connection with each individual student. The lecturer must make it perfectly clear that he is not directing enquiries to be undertaken by students on arrival. The general principles which emerge from the talk are, however, of importance in all work of appraising a milieu.)

SECRET

1. PROPAGANDA DEPENDS ON FACTS

    The propaganda worker must know the facts of the political situation. (Cf/ C.1). The propagandistic treatment of those facts is influenced by the opinions of the public to which propaganda is addressed. Methods of investigating opinions quickly and accurately form the subject of this talk.

2. DANGERS OF INADEQUATE INFORMATION

    Reports are constantly reaching us, often at the cost of agents lives, to the effect that "German propaganda is being successful"; that "The population of region X is 80% pro-British". These reports are valueless, because:

    a)   They do not yield facts precise enough to form a basis for work, and

    b)   They are guesses.

    What we need are reports of observed fact, the accuracy of which is above suspicion.

3.   INVESTIGATION AS A SCIENCE

The science of investigating opinion has made great strides in recent years, and the movement of large sums of money in commercial advertising and market research is actually determined by its findings. Cf. Fortune and Gallup Surveys; Literary Digest Straw Votes; Mass Observation.

Concrete cases of political and commercial propaganda based on good or bad interpretations of public opinion. (E.g. Petain and Lux soap.)

4.   Investigation methods in their normal peacetime thoroughness are obviously not applicable to our work. But the principles that govern it must also govern us, just as the propaganda principles which govern the vast output of the B.B.C. must also govern the small town printer working in a basement by candlelight.

SECRET

5.   "THE RANDOM SAMPLING METHOD"

This is the most accurate method of investigating opinion. Broadly it may be described as follows:

If a large group (Group "A") is under investigation, a sample (Sample "a") is taken. If this is accurately done, i.e., if Sample "a" is completely representative of group "A" the trends opinion found in the sample will be found in the group and in the same proportions. Similarly for a complicated group (A B C D) when sample a b c d will be required with "a" proportionate to "A" and "b" to "B", etc.

To achieve accuracy, rigid adherence to certain rules is required:

a) <u>A sample need not be large</u>.

   E.g. the Gallup forecast of Roosevelt's second election was based on only 2,000 views out of 130,000,000, yet it came within one percent of accuracy. For small town, local or craft reconnaissance, a much smaller number is sufficient.

   N.B. Whereas Gallup and other nationwide surveys organized on a commercial profit making basis require a high degree of accuracy, accuracy within 10 or 20 % is valuable for propaganda guidance.

NOTE: The introduction to Section 5 and also Para. i). can be omitted when students are lacking either command of English or minimal grasp of mathematics.

b) <u>The sample must be chosen at random</u>.

   Never let any arbitrary factor influence your choice. E.g. If we are investigating opinion in a coal mine, do <u>not</u> choose the foreman, the first ten men to reach the pithead, the secretaries of the local union, your brother-in-law, or anybody distinguished by unrepresentative characteristics.

c) <u>Check fact rather than opinion</u>.

   People are not reliable guides to their own opinions. In fact, on matters not violently affecting personal advantage, most people tend to give either the answer they think you want or an answer that does them credit. Peoples actions are a more reliable guide to their opinions than their words. E.g. Don't ask people their opinion on R.A.F. leaflets. Find out whether they pass them on.

d) <u>Use of observation rather than questioning</u>.

What papers do people read? Where do they shop? Do they talk to Germans? Do they watch German newsreels? Do they listen to the B.B.C.?

e) <u>Avoid all hypothetical questions</u>.

Instead of asking "What would you have done if you had not joined the army?", ask "What work were you doing or applying for when you were called up?"

f) <u>Avoid all words of vague meaning</u>.

Words like "often", "much", "occasionally" tend to obscure facts rather than reveal them. They open the door to all kinds of wishful thinking and guesswork.

g) <u>Never try to prompt people</u>.

When it is difficult to get an opinion, there is a temptation to suggest one. This is a frequent source of error. No answer is better than a prompted answer; no information better than misleading information.

6.

FOR REPORTS NEVER CONFUSE FACT WITH INTERPRETATION

Interpretation may be useful on occasion, but, when reporting, always distinguish clearly between fact and interpretation. Reports should read something like this:

"<u>Fact</u>: X miners and Y railwaymen in Blanktown have volunteered for work in Germany within the last Z days. Miners give reasons A, B, & C; railwaymen give reasons B, C, & D.

<u>Interpretation</u>:

On the basis of the above facts, my opinion is that Germany's attack on Russia has produced a marked swing towards collaboration among working-class circles in Blanktown."

(Students are asked to criticize this report for errors or omissions.)

On the basis of sound information, collected in the manner described above the formulation of effective operational propaganda is facilitated.

7. <u>SECURITY PRECAUTIONS</u>

SECRET

a) Ideally all questions addressed to a group vis-a-vis an identical political fact should themselves be identical.

With the principles enunciated above in mind, they should therefore be carefully phrased in such a manner that they can be slipped innocently into any conversation.

- 77 -

If this is impossible for security reasons, take care to phrase all your questions so that the answers "add up". E.g., the answer "Yes" to one question would be useless when juxtaposed with the answer "on a Wednesday" to another question on the same subject.

SECRET

b) <u>Behavior</u>

Apply methods of information gathering listed under A.3 Sec. 3) b).

Do not frequent too obviously certain areas or cultivate too obviously the members of a particular trade.

It is possible to elicit information without appearing to ask for it, to overhear gossip in cafes and queues, to observe behavior in the group affected. This is proved by the fact that people have done it.

H

SELECTION OF TARGETS

1. **DEFINITION**

    A target is anything of value to the war machine which can be attacked.

2. **POLICY**

    a). Extension of long-term bombing policy by destroying commodities of which the enemy is most in need and which cannot easily be bombed. In this connection, remember:

        i. Often you will receive directives from a higher source as to the most useful targets to attack.

        ii. Synchronization of such attacks adds to their weight. Over a wide area this can also be organized from above.

            <u>Examples</u>: Railways, shipping, canals, rubber, fuel-oil, lubricants.

    b). Constant irritant to inconvenience enemy by making his position more difficult, straining his C.E., lowering his morale and encouraging civilians. In this connection, remember:

        i. Select targets that harass enemy simultaneously in various parts of occupied territory. The wider the distribution of attacks the greater the strain on his C.E. thereby minimizing chances of detection.

    ii. Hammer at enemy incessantly. Saboteurs should be busy; enemy unhappy and constantly surprised.

    iii. Get at the enemy's vitals, e.g. industry, war production, food.

    iv. Get at his transport, rail, road, river; and at sea-going ports.

    v. Get at his communications; telegraph, telephone, wireless stations and postal services.

    vi. Liquidate his key personnel. Technicians, experts in local affairs. It is often better to attack a subsidiary part rather than the main target, particularly when it is a bottle-neck - e.g. instrument works supplying several aircraft factories.

c). While planning major disasters for the enemy, cause as much minor inconvenience to him as possible by:

    a). Petty acts of damage.
    b). Tyre damage to cars.
    c). Derailment of Trains.

Cultivate the sabotage eye; learn to look at things from the point of view of how to destroy them.

d). <u>Future possibilities</u>.

    a). Make continual contacts with people; you may not want them at the time; in the future they may be invaluable.

    b). Keep a tab on likely people's hobbies, strong points, weaknesses.

    c). Don't neglect targets because existing conditions may exclude action. An opportunity may occur later - e.g. invasion.

e). <u>Flexibility</u>.

Be ready and teach your agents to be ready to scrap plans that have perhaps taken months to prepare.

3. <u>GENERAL SURVEY</u>

The first step is to survey all possible targets for immediate or future attack. Put these in order of priority, having regard to directives from home. In order to do so consider the following points:

a). How far will successful attack affect the enemy?

b). How far will successful attack affect local population?

      c). The time factor; when will it have the greatest effect?

      d). How far can it be co-ordinated with other attacks, thereby increasing effectiveness?

      e). The strength of the forces at your disposal and the capacity of your operators.

      f). Weigh the cost of the operation - i.e. casualties, prisoners, reprisals, the chance of having part of your organization lost by reprisals for a small operation, thus prejudicing a larger one.

4. <u>ENEMY REPRISALS</u>

These are usually inevitable. Consider effect on public opinion in selecting targets. Will it turn local population against operators? On other hand morale is often stiffest where sternest reprisals are taken. They then usually cease to have deterrent effect (e.g. Poles).

Your main weapons against reprisals are:

      i. <u>Make damage appear accidental</u>.

          Choice of methods restricted. Do not fire a building not liable to fire; do not use explosives where there are none. First attempt must be successful. Duplicate apparatus to make sure that at least one works and no traces are left.

ii. <u>Exonerate local population</u>.

   Seize opportunity of a raid to:

   Make job look like bomb-damage.

   Make job look like long-distance shelling.

   Make job look like work of landing-party with whom locals have had no contact.

iii. <u>Choose time when enemy cannot avenge himself</u>.

   For instance, before invasion by enemy of another country. Damage to V.P.'s behind his lines will then have even more important results with no evil consequences to civilians.

I

PLANNING & METHODS OF ATTACK

1. **INTELLIGENCE.**

    The prelude to all operations is good information. This can be acquired by two methods:

        a). Informant service.

        b). Reconnaissance.

    A). It is from your Informant service that you will receive the information regarding possible targets to attack, e.g. industrial information from factory worker.

    B). Make your preliminary recce. on the following principles.

        1. Either determine to be seen or not to be seen. In each case have cover-story ready.

        2. Observe factors against you.-viz. number and location of sentries, alarms, locks, walls, barbed wire, general security measures.

        3. Observe factors in your favour. - viz. covered approaches, nature of ground near target (Noise factor).

        4. Estimate most favorable time and weather conditions for operation. Don't forget feast days.

5. Observe line of retreat. This, in irregular warfare, is more important than line of attack.

6. Estimate size of raiding-party.

7. To clarify operation orders if necessary, prepare map enlargement of area and sketch-plan of building. Destroy as soon as possible afterwards.

2. <u>FINDING THE WEAK SPOT</u>.

For our purpose it is often best to attack one weak portion of the target than the whole target, thus immobilizing the whole target, e.g.

    a). Citroen works making German tanks, destroy plugleads which are made in small subsidiary factory.

    b). Condensed milk factory - destroy supply.

    c). Aeroplane factory - destroy instrument supplies. (In other words, go for the bottle neck.)

3. <u>WEAPONS</u>.

Two main types:

    a). <u>Explosives of all kinds</u>.

        These naturally require

specialist training and involve questions of supply and transport.

    b). <u>Natural elements</u>.

        Full use should be made of:

1. Fire
2. Water
3. Wind & draughts
4. Sand, iron filings, etc.

4. <u>OPERATION ORDERS</u>.

Draw these up (mentally?) under the following heads:

    a). Object of attack
    b). Information
    c). Method
    d). Treatment of casualties, line of retreat & rendezvous
    e). Whistles, call-signs, passwords, etc.

5. <u>PERSONNEL</u>.

Organize your personnel as follows:

    a). If necessary, form a special group.

    b). Try to enlist persons with "inside" knowledge and with a high reputation in enemy eyes - or at least not suspected.

    c). Do not overlook the possibility of using women.

    d). Keep members unknown to each other for as long as possible.

    e). Issue detailed instructions to each member either individually before the operation, or collectively as late as possible before zero-hour. Delay revealing zero-hour as long as possible.

    f). Supply everyone with a story to cover as much of his activity.

    g). Carefully organize getaway for each person. Make arrangements for wounded not to fall into enemy hands.

    h). Do not forget food and water for operators before job.

6. **AFTER ATTACK**.

    Observe security principles - forbid association, celebrations, tell-tale signs.

IMFORMANT SERVICE

1. **INTRODUCTION**.

    Without good information it is impossible:

    a). To protect oneself against enemy C.E.

    b). To plan or time operations. (c.f.

importance of "I" to "O" staff in regular army.

2. **WHAT DO YOU REQUIRE TO KNOW?**

    a). Local Conditions -

    - Unprocurable articles (e.g. danger of ordering wrong drinks or cigarettes.)

    - Transport service (e.g. fewer trains, buses, taxis) and restrictions (e.g. reason for travelling.)

    - Market days. Danger of search for "Black Market" goods.

    - New slang or colloquialisms brought about by war.

    - General temper of local population.

    b). Local Regulations -

    - Identity papers. Are yours in order? (Compare yours with other people's and, if possible, procure genuine ones.)

    - Ration cards. Find out how to procure these.

    Movement restrictions. What passes are necessary?

    - Control posts. Manned by enemy troops or local police?

- Evacuation from forbidden zones.

- Curfew hours.

- Blackout regulations.

- Bicycles, - licenses, restrictions, etc.

c). Enemy Methods and Personnel -

- Location of troops.

- Location of nearest enemy police or Gestapo, with details or personnel; attitude of local police.

- Names of civilian police spies, agents provocateurs.

d). Operational Information -

- Possible targets: Enemy communications, H.Q., dumps, factories.

- Bottle necks in enemy production and communications.

- Internal working of factories, power station, railways, etc., e.g. type of machinery used.

- Personnel employed in any of the above.

- Means of entry: layout, guards, control system.

- Documents: workers passes, blueprints, etc.

3. <u>HOW DO YOU OBTAIN THIS INFORMATION</u>?

    a). By direct interrogation.
    b). By constant personal observation.
    c). By reading newspapers and listening to radio.
    d). By Informant Service.

    THE INFORMANT SERVICE:

    a). <u>Personnel</u>.

        (i) Very few should know that they are informants. The great bulk will be quite unconscious of it.

        (ii) Select from as many strata of society, trades, professions, etc., as possible.

        (iii) Best are types who constantly mix with all sorts. - E.G.,
          Priests,
          Inn-keepers,
          Waitresses, barmaids,
          Doctors, dentists, hospital staffs,
          Postman, telephone & telegraph operators,
          Bankers, shopkeepers,
          Railways officials & workers
          Servants,
          All grumblers & malcontents.

        (iv) In due course you may decide to approach a few

of the more trustworthy informants with a view to recruiting them.

    b). <u>Methods</u>.

        (i) Journalists technique of eavesdropping on the masses. c.f. - Ability to hear and separate two simultaneous conversations while ostensibly listening to a third.

        (ii) Taking advantage of other people's bad security - e.g.

- Careless talk.
- Disgruntled enemy personnel.
- Affecting ignorance & thus encouraging others to air their knowledge.
- Making false statements to elicit correct reply.

        (iii) Do not discourage informants, however trivial the information. c.f. - reporter's maxim: "Never refuse a date".

HANDLING OF AGENTS

1. <u>FINANCE</u>.

Once an agent is recruited:

a). Put the matter immediately on a business footing. Settle financial side first.

b). Consider grading your agents from 111 to 1 on a rising scale of salary - to promote keenness.

c). Differentiate between salary and expenses. If a patriot wishes to work for expenses only see that he is not out of pocket.

d). Pay punctually the exact amount.

e). If you find you cannot pay on promised date, say so at once, and state when you think you will be able to.

f). Determine to have no money disputes with your agents. If, after careful inquiry, you are certain that agent is exceeding his necessary expenditure, pay him the first time and ensure that subsequent excesses are discontinued.

g). Help agents who fall ill or get into difficulties as would a good employer. Pay dependants of those under arrest or pensions to families of those liquidated.

h). There is only one way to count the cost in this work; balance the total expenditure on your agent (pay plus expenses) against value received.

i). Payments, particularly for irregular work are sometimes better made in kind or in services - e.g.

   (i) Special foods which are unobtainable - sugar, chocolate, soap, milk, cream, coffee, tea, cigarettes, tobaccos, spirits, wine, beer, petrol, razor blades, silk stockings, ladies' clothing.

(ii) Obtaining employment for agent or his family or friends.

(iii) Business or social introductions.

(iv) Medical assistance or supplies - e.g. Insulin for diabetics.

Note: The uses of blackmail should not be overlooked, e.g. recruiting a German, ensuring silence, obtaining a pass or special facilities.

2. <u>SECURITY</u>.

a). He must continue in his present job or another. Cover comes first.

b). Test your newly recruited agent on a small job before you entrust him with any real work. Check up on his report; this will reveal his efficiency as well as his reliability.

c). Only employ one agent on one job. He should know nothing whatever about other sections of your work.

d). An Agent's knowledge of your organization should be strictly limited to the persons and information which are essential to the performance of his particular job.

- e). Avoid numbering agents (most dangerous - viz. German agents) or use of aliases. Christian names are safest.

3. <u>TRAINING OF AGENTS</u>.

    - a). Necessity to train all agents in principles of security and cover first.

    - b). They should receive sufficient technical training to enable them to carry out their job.

    - c). They should not receive training in other branches which do not concern them.

    - d). Question of security during actual training, e.g. skiing expeditions in Norway, Country houses and hiking parties. Stress danger of regular attendances of several persons at one center.

4. <u>DISCIPLINE</u>.

    - a). An agent may become inefficient - either through overwork and nerves, or through being won over by the enemy. Lay off gently; go on paying salary promised; have him watched carefully. If he is overworked he may improve. If he has gone over, he may give his hand away. If certain that he is a traitor, kill him. (You may have to sacrifice a valuable man to do the killing, unless it can be made to look like an accident.)

    - b). "Semi-honest" agents may, directly or indirectly, try blackmail. The

clever one will say "I am being
blackmailed. Help me." Investigate
his story in detail. He may be
speaking the truth. If not, unless
you can successfully put the fear of
God into him, you may have to kill
him; or you may be able to convert
him into a double agent.

c). The greatest safeguard against slack-
ness, blackmail and all forms of
indiscipline is to foster in your
agents the idea that they are working
for an immensely powerful organization that can be benevolent to loyalists and ruthless with rebels.

(No one dared blackmail the Mafia.)

d). Make every-one believe that if you or
they are killed by the enemy, there
will be others to take your places.
The enemy cannot break your spirit.
Even if part of the Organization is
betrayed, the rest will remain and
surely avenge itself on the traitor.

OBSERVATION & SURVEILLANCE

1. **INTRODUCTION**.

    Observation is the power to observe.
    Surveillance the method of carrying out
    observation for a set purpose.

2. **GENERAL**.

    1. Observation means that one should
       look at an object not as any
       absent minded passer-by but with
       penetration.

2. In "every-day life" observation increases interest in things going on around one.

3. In "our business" observation is one of the agent's main Lines of Protection.

4. The power to observe is found in some people as a gift. When you hear such people talk you are struck by the amount an unobservant person misses. Other people are unable to use this power of observation even sufficiently to preserve their own simple interests. How much less, then, are they to do it for the sake of saving their own lives.

5. Good observation implies not only seeing but hearing, feeling, smelling and awareness of what is going on around one and why.

3. <u>METHOD OF IMPROVING POWERS OF OBSERVATION</u>.

This will mean hard work, time, interest, desire to succeed and systematic practice.

1. First take things that please you - please your eyes, your touch, your sense of smell and hearing. Then try and express in words what it is in these things that pleases your senses. Such an effort will cause you to observe the object more closely in order to appreciate it and find its qualities, e.g. a girl, a building, a flower, a scent, a voice.

2. Take things you dislike and repeat the same process.

3. Take things that to you are neutral and repeat the process, analyzing towards it.

4. Then turn to what the human race has to offer in art, music, literature, architecture, etc.

5. Study the impression you get from direct personal contact with other people; put into words: the facial expression, the eyes, the voice, the gestures and the speech of people and report process, asking yourself if you like them or dislike them, and why.

6. Then study the actions of people, helping yourself with the questions: "Why did he do that?" and "What did he have in his mind?" thus practicing powers of deduction.

7. Then try and judge character from clothes, such as the dandy with his hat on one side, peculiarities of dress, walk, etc. Allow natural inquisitiveness of the human being to have full scope. Have a healthy suspicion. At first this will be a strain but in a very short time it will become a habit.

8. Observation must be systematic. Train yourself to follow a regular routine in observing persons, places, etc.

9. Thus the agent in his own locality by his powers of observation will know most people by sight - what they are, what they do, why. He

will know what type of person he has to deal with. He will also know other details, such as houses, shops, traffic and other objects.

10. By this means presence of strangers or unusual happenings is detected. Thus control of identity is begun and supervision established.

4. <u>OBSERVATION AND SURVEILLANCE</u>.

The subject of observation, surveillance and description are so closely allied that one is of no particular value without being supplemented by the other two. Surveillance would be of no value without the observation of specific objectives and the lack of ability to report your findings from observation and surveillance, i.e., description, would render your first efforts futile.

<u>OBSERVATION EXERCISES</u>

<u>OBJECT</u>.

To develop and train the powers of observation and memory on the part of the Students.

a). <u>Kim's Game</u>.

The Instructor places about 15 or 20 different articles on a tray or table, e.g. two or three different kinds of buttons, pencil, cork, stones, knife or anything else handy. Then covers them up with a cloth. Make a list of the articles, and put columns opposite these for each of the students replies. Then uncover the articles for a minute, look them over again taking each student separately and getting him to repeat as many of the articles as he can remember marking off each item against his name on the sheet.

<u>Variations</u>.

1. Change position of objects on the tray so that the Student has to memorize those changes of position as well as the actual objects.

2. Students to memorize the objects in groups such as articles of stationery, etc.

3. Incongruities might be introduced, e.g. a pipe with an unused bowl, yet a well bitten stem. Match box containing wrong type of matches. An ink bottle marked "blue ink" containing red.

b). <u>Examining an object and writing a description of it</u>.

This exercise is similar to (a) except that a single and more complicated object is used, such as a photograph or picture depicting some activity. Students should be asked to give a full description of everything seen, position of persons in a picture, what they are doing, clothes, etc.

Objects suitable for this exercise might include photographs and pictures of activity, drawings, etc. of uniforms, e.g. foreign police, labour corps, etc. Different types of civilian clothing, a front page of a newspaper, etc., etc.

c). <u>Memorizing a newspaper in a short time</u>.

Imagine that the student is a person who has been left alone in an army or police office for a period of a minute or so, the student should study a document and memorize the essential points. Documents giving lists of names, instructions or details of stores, equipment, etc., details of a new identity or ration card, to be introduced shortly, might be used.

d). <u>Memorizing a map</u>.

It is essential to use only a portion of a map, which shows a few salient points, rather than large and complicated picture. For instance, a portion of a large-scale map of this district, showing the towns of Southampton, Lyndhurst, Brockenhurst, Lymington, etc. is a good example.

<u>Variation</u>.

Memorizing a route between two fairly distant points after a short study of the map. The journey to be undertaken later.

e). <u>Identifying objects by touch</u>.

In this case the students must be blindfolded. They can either name the objects in turn as they touch them or memorize them all at the end of the exercise.

This exercise can include articles that might be required for night operations, i.e. ammunition, plastic, detonators, adhesive tape, different types of fuses.

f). <u>Identifying objects by smell</u>.
Similar to above, using articles recognizable only by smell, e.g. tobacco, vinegar, paraffin, ink, boot-polish, brasso, etc.

g). <u>Identifying sounds</u>.

There are two forms of this:

    a). Identifying the nature of certain sounds on the lines of a B.B.C. Broadcast, E.G. a window opening or shutting, man in boots, gym shoes, wellingtons, walking on gravel, etc., etc.

    b). Identifying, when blindfolded, the direction from which a sound (e.g. ticking of a watch) proceeds.

h). <u>Observation of surroundings</u>.

Students will be asked to describe various details of the house in which they are living, e.g. number of stairs, number of panes in their bedroom window, colours of walls and furnishings, tiles, etc., also means of entry or exit in emergency.

i). <u>Plans by memory</u>.

Students will be asked to produce without notice a plan or elevation of their house or one of its rooms, or a sketch plan of the garden, and to give particular details of flowers, trees, outbuildings, etc.

j). <u>Noting changes</u>.

Students will be brought into a room which they have temporarily evacuated and asked to describe various changes which have been made in the room during their absence.

k). <u>Reconstructing a burglary</u>.

Students will be asked to note and deduce from signs left by a marauder entering their premises, e.g. footmarks, trodden grass, broken twigs, signs of tampering with windows, etc. They should be asked to reconstruct the man's actions from the signs they have observed, and any signs not observed should be pointed out.

l). <u>Deduction of a man's profession</u>.

This is to be done entirely from the contents of his pockets, which are displayed, e.g. letters, diary, bills, bus ticket, visiting cards, etc.

m). <u>Characteristics of trades and professions</u>.

Students are to write down the leading characteristics, appearance, dress, mannerisms, habits, etc. of a number of trades and professions.

CLOTHES SEARCH

<u>Object</u>:

(i) To train students to conceal small objects or papers about their person.

(ii) To give practise in body searches.

(iii) To stand a search without betraying the hiding-place by reactions.

One student will conceal a message or object in his clothes. There will be no time limit for this operation.

Two students will play the part of searchers. No interrogation will take place during the search, which should be conducted as follows:

The clothes will all be placed on the table. One searcher will then proceed to examine them, one article at a time, in each case placing the article after examination on a second table. The second searcher will watch the suspect continuously for any signs of reaction as each article is dealt with - e.g. anxiety or relief.

HOUSE SEARCHES & RAIDS

1. <u>INTRODUCTION</u>.

  a). Lecture studies police method, i.e. defensive point of view, but incidentally provides guide for planning of offensive raids by our organizations.

  b). Police searches and raids undertaken with any of following objects. Locating -

    (i) Individuals.
    (ii) Documentary evidence, propaganda.
    (iii) Arms, explosives.

  c). May be

    (i) Routine search (cf. snap control raids on night clubs.)
    (ii) Special search, based on evidence or suspicion.

      d). May be conducted

          (i) In agent's presence to effect arrest.
          (ii) In agent's absence to confirm suspicions and collect evidence.

2. <u>PREPARATION OF SEARCH</u>.

    Police will make detailed preparations in advance element of surprise essential - unsuccessful search only arouses suspicions. Following are guiding principles:

    a). <u>Reconnaissance</u> - Plan of house (lay out of rooms, exits - provide check on possible hidden rooms, c.f. Libre Belgique) approaches, ground surfaces, walls, trees, flower beds, garage, outhouses, A.R.P. shelters, coal chutes, well, telephone. Recce. must be observed.

    b). <u>Occupants</u> - Their habits. (as guide to suitable time for raid) Servants (may be used as informants, c.f. burglaries with "inside" help* dogs. Use of local informant service as guide.

    c). <u>Hour</u> of raid according to nature (cf. 1 (d)).

    d). <u>Password for party</u>.

3. <u>ORGANIZATION OF RAIDING PARTY</u>.

    Numbers according to circumstances but in addition to searches - cordon or guard for exits may be required.

4. **METHOD**.

   a). Approach must be unobserved or sudden descent.

   b). Guard exits and/or throw cordon round building.

   c). Simultaneous penetration at all entrances - speed in taking up posts essential.

   d). Search for individuals. Muster in one room, disarm and search. No smoking, eating, etc.

   e). "Wanted men" taken to own rooms, which are searched in their presence. Ownership of articles established.

   f). Search remainder of house.

   > NOTE: Some evidence of Gestapo method, using four or five men, each searcher accompanying 1 or 2 occupants during search.

5. **POSSIBLE HIDING PLACES**.

   a). Permanent.
   b). Emergency.

   1. **Persons**.

      On or under roof, chimney stacks or cistern tanks, behind curtains, doors, false doors; inside wardrobes, clothes baskets, dustbins; between ceilings and floors; under stairs, in false tops above cupboards, in lofts, under coal in cellars, in trees and dugouts.

2. <u>Documents</u>.

Behind pictures; under sides of tables and carpets - in fireplaces, hollow beams, holes bored in tops of doors, in furniture, windows, behind skirtings and picture-rails.

3. <u>Arms and explosives</u>.

Under refuse heaps or litter, in watertight cases in tanks or wells; cartridges in holes in woodwork - loose bricks, hollow piping, etc.

(N.B. Use of simplicity as opposed to extreme subtlety - eye level.)

6. <u>PROTECTION AGAINST SEARCH</u>.

a). Effective concealment may defeat <u>routine</u> search but if possible keep no incrimination material "at home" particularly when under suspicion. Possible use of abandoned houses or apartments.

b). Practise orderliness - search during absence must not pass unnoticed.

c). Informant in police circles to warn of impending search - note German habit of searching all houses in a street or quarter after demonstrations or unrests.

d). System for warning off collaborators when house under observation.

e). Particular danger - Gestapo waiting in house until agent arrives, subsequently arresting all callers and breaking down organization. (e.g. arrested couriers forced to reveal "boites aux lettres".)

7. <u>OFFENSIVE RAIDS BY OUR ORGANIZATIONS</u>.

e.g. to "liquidate" Agents provocateurs, destroy H.Q.'s, etc.

a). Main principles of preparation still apply.

b). Individual function of each member and general plan well rehearsed.

c). Study line of retreat.

d). Prearranged emergency signal and R.V.

e). How to deal with guards - evade or overpower.

f). Note types of guards employed.

1. Sentry at gate, remainder of guard within call. Possible R.P. or C.M.P. support at H.Q.'s. At night covered by second sentry, invisible but within firing distance.

2. Stationary sentry concealed, commanding view of approaches.

3. M.G. or L.M.G. post on "fixed line".

4. Guard on regular beat.

5. Flying patrol around perimeter often at irregular intervals.

6. Chain patrol - 1 man being dropped off at selected intervals and remaining stationary - very deceptive on noisy nights.

7. Picquet with lines.

8. Any other permanent defense posts - P.A.D. Fire Picquet, L.M.G., A.A. positions permanently manned.

BODY SEARCH

1. GENERAL.

   a). Agent, particularly courier, sometimes forced to carry documents on person - not all messages, plans, etc., can or should be memorized.

   b). Enemy C.E. aware of this and use Body Searches as detective measure.

2. TYPES OF SEARCH.

   a). Routine: e.g. at Ports, Frontiers, Protected Area Control Posts - measure against smuggling goods, arms, currency. Impossible for authorities to search thoroughly everybody passing the control; danger of snap control (thorough search in selected cases - e.g. searches of munition workers at Ordnance factories).

b). <u>Special</u>: In cases of suspicion, police may search detained or arrested suspect prior to interrogation. When documentary evidence is sought, search will last many hours (CF. C.I.D. 2 days).

3. <u>POLICE METHOD</u>.

   Lecture considers procedure for thorough detailed search. Police will search for specific purpose e.g. arms, message, incriminating correspondence, currency.

   a). <u>Before search</u>.

   1. Police will try to catch suspect unaware - e.g. by sudden arrest, or by detaining ostensibly on "routine" business.

   2. Close supervision immediately before search - no eating, drinking, smoking, communication with other persons. (Possibility of suspect "planting" evidence elsewhere.)

   3. Police will ensure that suspect has not left behind any article of clothing or baggage (coat, valise, newspaper, umbrella.)

   4. For legal reasons may establish ownership of personal effects.

   5. Suspect asked to produce any documents - undeclared docu-

ments discovered during search incriminating per se.

b). <u>During search</u>.

1. Second witness may be used to observe facial reactions of suspect.

2. Interrogation not carried out at the same time but conversation may be used to trap suspect into an indiscreet statement.

3. Baggage and effects dealt with separately.

4. Search will be methodical, first clothes, then body, working from head to foot.

c). <u>Places of concealment</u>.

These are infinite - e.g.,

1. <u>Clothes</u>: Hat band, lining, collar, tie, tiepin, shirt, (under trade mark), lapels, padding in shoulders, double pockets, seams, buttons, soles and heels of shoes, tags of laces.

2. <u>Body</u>: Hair, nose, ears, mouth, stopped teeth, false teeth, glass eye, finger and toe nails, bandages, writing on skin.

3. <u>Effects</u>: Glasses, lighted pipes, cigarettes, sticks,

umbrellas, split postcards, chocolate, envelopes, books, newspapers, hotel labels, camera, invisible ink in ointments, soap, brilliantine.

d). <u>After search</u>.

Suspect not allowed to regain contact with unsearched persons or go back to room occupied prior to search - careful surveillance may be exercised for considerable period.

4. <u>THE AGENT</u>.

a). Avoid carrying incriminating documents when possible.

b). If selected for search at a control, don't assume the worst; search may be routine "frisk", carried out by inefficient officials.

c). Effectiveness of search depends on human element, e.g., N.C.O. searcher unwilling to take responsibility for ruining clothes, without specific instructions.

d). Decide what the object of the search is - arms, currency, documents, etc., e.g. may be able to focus attention on money and thus safeguard messages.

e). Watch your reactions - attitude consistent with cover - e.g. indignation, resigned dignity, willing collaboration - no visible relief when search is unsuccessful.

f).   Try bluff - direct searcher's suspicions to innocent article by simulating consternation and thus distracting attention from actual evidence cf. Louise de Bottignies and the lace.)

WIRE TAPPING IN OCCUPIED COUNTRIES

Numerous reports have come from occupied countries telling of the cutting of telephone wires that are being used by the Axis authorities. This practise is dangerous, being punishable by death, and in general can only result in minor inconvenience and delay to those in control. It has occurred to _____ that the activity of the patriots along these lines might be used to greater advantage. The following is an outline of his suggestions and of further information which I obtained from an expert in technical matters concerned.

If wires can be cut, can they not also be tapped? The risk would be increased because the agent would need to stay near the wires for considerable periods of time, but decreased because the users of the line need not know that it was being tampered with. The advantages to us would be very great if information gathered by tapping could be transmitted promptly to us, and this in many cases should be possible. The necessary equipment, which is small and light, could be carried into occupied countries by our agents, and could sometimes be obtained on the spot.

A second possibility is to use the threat of wire tapping as a psychological weapon. We might, for instance, arrange to broadcast, from suitable stations and at regular intervals, warning about the unnecessary risk of wire cutting and instructions in the method of tapping. This would give he local Axis garrisons and higher authorities something to worry about and might result in their drawing off extra troops for useless patrol duty, even if no tapping were actually accomplished, or attempted. In order to disturb the mind of the enemy we should, of course, have to send out accurate instructions in genuinely workable methods which would employ equipment available to the local population.

In a conversation on _____ with _____ , _____ for the _____, I learned that the question of equipment is no obstacle to the carrying out of either plan.

Tapping a telephone circuit is a simple process. Any telephone receiver hooked across the circuit, with a fixed condenser of 0.1 to 0.5 Mfd. capacity intervening above or below the receiver, will allow the listener to overhear conversations on that circuit. The following diagrams illustrate graphically the arrangement to be used with one-wire and two-wire circuits:

One wire circuit.

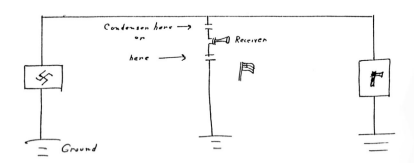

Two wire circuit

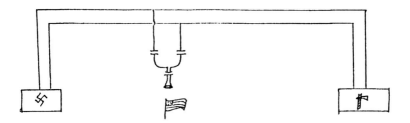

The receiver of a hand-set or "French" telephone can be used. In that type of instrument there are three wires: one to the receiver, one to the mouthpiece, and one common to both. If the wire tapper were not familiar with the construction of the instrument he would need to find the right pair by experiment.

If a condenser of measured capacity cannot be obtained, a working substitute can be easily made by binding wires to two flat conductors which are separated by a non-conductor; for example, two coins or flattened cartridge cases separated by a dry piece of paper.

Ground connection may be made by driving a metal rod, spade, pickaxe, or the like into the earth or by submerging a metal bar, plate or length of exposed wire in a lake or stream, or in the Sea.

The drawback to this simple system is that the tampering can be detected by the operator at either end of the line if he is on the alert for it. To obviate this danger, _____ has perfected a small device (smaller than a man's fist) which, through the turning of single knob, masks the tapping from the operator. _____

has some five hundred of these devices on hand and not in service, and we could have almost any number of them for the asking. Letters of request should be addressed to:
_____ and they should be specific, e.g. "We request the loan of 10 telephone monitoring handsets for an indefinite period...."

If it is decided that we should use the first plan (introducing equipment in order to gather information), it will be wise to make request for a dozen or more of the "monitoring" sets in the near future, in order to have them ready for service at short notice. Other branches of SA/B and SA/G should also be informed of the availability of the equipment.

If the second plan (threat of tapping as a psychological weapon) is adopted, the matter is taken out of our hands.

J

PASSIVE RESISTANCE

1. <u>GENERAL</u>.

   In enemy territory the civilian population may become a formidable problem. Concerted action is required. Organizations should be formed to educate the people to become a general nuisance to the enemy.

2. <u>OBJECTS OF PASSIVE RESISTANCE</u>.

   - a). To reduce enemy war effort.
   - b). To make enemy's life uncomfortable.
   - c). To raise the morale of civilians.

3. <u>PRINCIPLES OF PASSIVE RESISTANCE</u>.

   Civilians must learn to arrange their habits to cause inconvenience to enemy in every sphere. This is contrary to their natural instincts. They must be taught ways of doing this without risk of punishment. If passive resistance results in reprisals the principle will not be widely nor enthusiastically adopted. Actions must be organized which everyone can undertake without risk.

4. <u>GERMAN WEAKNESSES</u>.

   - a). Urgent need for increased war production and transport facilities.

   - b). Shortage of police officials, foremen, etc.

        c). Lack of knowledge of language and understanding of people.

        d). Long absence and distance from home, uncertainty of news, etc.

        e). Vanity, self-consciousness, lack of humour, inferiority complex and expectation of easy life.

5. <u>METHODS OF PASSIVE RESISTANCE</u>.

        a). Workers can hamper enemy's war effort by causing waste and delay, by excess of zeal, by over-caution, by multiplication of small delays and errors by feigning physical incapacity.

        Insist upon unnecessarily high quality.
        Specify needlessly high standard.
        Reject everything imperfect.
        Do jobs too well.
        Insist on pedantic adherence to regulations.
        Demand written orders.
        Ask for frequent renewals and overhauls of machinery, "So as to produce best results."
        Use "weather" excuse.
        Aim at slight delay at every change of operation.
        Use 20% more oil on each machine and lubricate 20% less.
        Claim exhaustion from air raids.
        Work slowly and make mistakes as a result.

Use every opportunity to waste time demonstrations and deputations about "reasonable" complaints, e.g. inadequate P.A.D.

b). Citizens can create extra work and inconvenience for German officials by excessive zeal, over-caution, well-meaning stupidity, over-politeness, etc.

Constantly report "suspects", sabotage, etc. The police must follow up every such report.
Ask unnecessary questions about new regulations.
Frequently report to the police.
Fill in forms incorrectly.
Misunderstand instructions.
Helpfully get in the way of A.R.P. personnel during raids.
Be verbose and unnecessarily gratefully loquacious.

c). Citizens can make Germans feel uncomfortable by unfriendliness, tactlessness, "amiable" gossip, demonstrations of passive unity, etc.
Whisper and laugh in their presence, but have an explanation.
Don't hear what they say.
Don't see them.
Leave the room when they enter.
Condole with them about bad news and bombing of their

families, etc.
Spread scandals about improprieties and political suspicions.
Mix married women's names.
Refuse to buy German controlled newspapers or attend enemy entertainments.

6. INCENTIVES AND ORGANIZATION.

Passive resisters must understand that they are as important as saboteurs. They must have a yard-stick to measure the results of their efforts. This will maintain their enthusiasm.

Examples:

If foreign metal workers in Germany waste 5 minutes a day for 6 weeks they will have done the equivalent of sabotaging a locomotive.

By wasting 4 hours in a month metal workers in Occupied Territory will have achieved the equivalent of sinking a U-boat.

Recommend to each person or group only two or three things to do, so that they become expert. These things must be "safe". Such acts can be repeated without fear of detection. Nevertheless warn passive resisters not to go "too far" and to change their methods before the authorities have time to detect them.

7. More detailed methods of conducting passive resistance, catalogued by trades and professions, are found in the pamphlet which will now be distributed to you. You will also have fresh instruction in passive resistance for railway workers.

<u>GUERILLA WARFARE</u>

1. Main object of guerilla warfare is to harass the enemy both in his own country and in occupied territory.

2. <u>Effects of successful guerilla warfare</u>:

    a). Compel enemy to disperse his forces.
    b). Thereby weaken main armies.
    c). Demoralize his detachments protecting supplies.
    d). Reawaken civilian morale - passive resistance.

3. <u>Main types of guerilla warfare</u>.

    a). <u>Sabotage and Para-military</u>.

        1. Individuals or small groups working by stealth on acts of sabotage.

        2. Larger groups working as band under a leader and often employing military tactics and weapons.

        3. Larger guerilla forces with para-military organization.

            (We are still mostly in the

first stage, although the second has been reached some time ago in Poland. Successes, coupled with progressive demoralization of enemy troops, will lead to the third stage.)

    b). <u>Fifth Column activities</u>.

Propaganda, subversion of enemy's morale, ridicule (and finally destruction) of puppet regimes, various Fifth Column activities.

4. <u>Vulnerability of</u>:

    a). Enemy L. of C.
    b). Enemy transport and supply systems generally.
    c). Morale of enemy troops far from the homeland.

5. <u>Some essential factors of successful action</u>:

    a). Surprise, surprise, surprise. Hit hard where the enemy is most vulnerable, disappear completely and reappear for another raid somewhere else.

    b). Only engage in an operation which promises a 75% chance of success. Plan each operation with the greatest care; neglect no detail. Leave a reasonably safe line of retreat.

    c). Never carry incriminating documents on your person or leave them where they can be found.

6. <u>Need for leadership</u>. A leader:

   a). Must have courage and resource.
   b). Must be intelligent and capable of quick decision.
   c). Must know intimately the country in which he is operating.
   d). Must be able to use a compass and map.
   e). Must inspire his men with confidence.

7. <u>Qualities of leadership</u>.

   a). Capacity to view the whole situation calmly.
   b). Untiring search for enemy's weakest point.
   c). One cannot control others without being able to control oneself.
   d). His ideas must be simple and clear.
   e). A sense of what is possible with the forces at his disposal.
   f). Must know how to organize his time - a time for everything.

   <u>A leader performs four acts</u>:

   a). He inspects.
   b). He receives reports.
   c). He makes plans.
   d). He issues definite orders.

8. <u>Qualities of the guerilla soldier</u>.

   a). Must have greater general knowledge of offensive means.
   b). Must be able to select targets most vital to enemy.
   c). Must know how to time his blows.

9. Guerilla warfare is one of the great war-

winning weapons; virtual impossibility for enemy to combat subversive movements on a large scale simultaneously in all his conquered territories.

PLANNING OF SUBVERSIVE ORGANIZATIONS

1. INTRODUCTORY.

   It is evident that there are considerable difficulties in seeking to plan large scale subversive organizations in the middle of a world war. Unlike the Germans, who had their Auslandsbureaux organized in every important country of the world years before the war under the aegis of their diplomatic service and with no real difficulties of travel and other restrictions to contend with, we are now having to build up practically from zero with an enemy C.E. organization on tip-toe to discover any traces of British activities. The fact, however, remains that the Germans succeeded in building up their organizations in countries which were, at all events, latently hostile to themselves whereas we should be able to count on the active support of at least a substantial proportion of the population in those countries who want to free themselves from German domination.

2. METHODS & WEAPONS.

   To-day, those engaged in subversive work have a greater choice of methods and weapons than ever before. Also we have every modern invention to help us. If we consider methods first, we have:

   1. Political subversion directed to the overthrow of all puppet regimes.

2. Propaganda directed to harm and ultimately to overthrow the enemy.

3. Sabotage of all kinds.

4. Passive resistance.

5. Raids of all types.

6. Guerilla warfare.

7. Political assassinations and murder of enemy personnel.

8. Armed mass revolt.

Our weapons are innumerable. To mention only a few -

1. Political subversion and propaganda, written, broadcast, whispered.

2. Sabotage both with and without explosives.

3. Incendiarism.

4. Passive resistance, scientifically applied.

5. All forms of normal fire arms, bombs, knives, clubs and thuggery.

6. Poisons and drugs.

You can probably think of many more.

The means which we can employ to help us are unlimited and far greater than ever before. In addition to all the old means, we have -

1. The aeroplane: To drop our men and their equipment.
   To pick them up.
   To receive their signals.
   To assist them by bombing raids.

2. The wireless set: To communicate.
   To receive information.
   To give instructions.

3. The special sets to communicate with the air.

4. The broadcasting station for propaganda and communication.

5. Soluble paper, edible paper and silk paper on which to write.

6. Scientifically designed secret inks and ciphers to write in.

7. The submarine and the folding boat to put our men ashore and take them off.

8. The high speed launch to do the same.

9. Later, perhaps, the armoured landing craft.

I have mentioned all these things because I do think that all those who are planning for a country section will have to know a good deal about all of them. I do not say that they can or should be technical experts, but that they should know the powers and limitations of all these weapons and means so that they can appreciate the part which they can play in their schemes and use them to their best advantage.

3. THE GENERAL PLAN.

The department, presumably, has, or has now under preparation, a general or master plan on the most comprehensive scale coordinating the world wide activities of its country sections in the various spheres of sabotage, political subversion, propaganda and guerilla warfare to the one object of defeating the enemy. This plan is no doubt the master plan to which all minor operations are related.

It is, of course, very unlikely that the whole plan, or any substantial part of it, can be put into operation at an early date, but this general plan will give the indication of the line on which each section works.

4. THE SUBDIVISION OF THE PLAN.

The general plan is, then, presumably divided into sections and portions of it are entrusted to the country sections concerned to put into operation in the same way that the C.-in-C.'s plan is sectionalized to Corps, Divisions and Brigades, each of whom carry out their share. Individual operations will probably only be a minute part of the general plan, but this is inevitable and immaterial so long as they conform to that plan.

THE COUNTRY SECTION AND THE PLAN.

On receiving their part of the general plan, no doubt Country Sections study it to determine how best they can give effect to it.

It is impossible to go into detail, but in general principle this will probably mean dividing the country up into a number of zones. Each of these zones should be under the control of a zone organizer. These zone organizers should not be known to each other.

Unless very exceptional men are obtainable it will often be necessary to recruit a different type of organizer for each type of work: propaganda, sabotage, guerilla warfare, etc.

It is not a good thing to be hypnotized by zones as drawn on maps. Sometimes it may be better, especially in small countries, to divide the country into categories, either religious, social or professional and organize on that basis, e.g. railway workers, seamen, civil servants, religious orders.

The whole plan must be elastic and capable of modification. We have to work with the men we can get and it is not always possible to get a man suited both for the type of work required and the zone in which he is wanted. Also at times you may find an exceptional man around whom it is possible to build up a whole section of the organization, thus modifying the original plan. When making a plan we should certainly take account of the various underground organizations which already exist in occupied territory. Nearly all occupied countries have these organizations, but there is a danger of their fading out for lack of encouragement and financial support. For reasons of security these should be contacted through cut-outs. We know of their existence in Norway, France and Belgium and also in Holland.

Their full extent is not known, but they certainly seem to be fairly widespread. If their activities are sporadic and ill timed, they may be an embarrassment to us so that not only do they provide a valuable basis on which we can build, but also their activities should be coordinated with our general plan.

We should not neglect minorities living in occupied countries. These may be a great help, e.g. Poles, Spaniards, Italians. Also we should start now to set up an organization in all countries which are not yet occupied but well may be, i.e. Turkey, Persia and Afghanistan.

5. METHODS OF WORKING THE PLAN.

While dividing the area into zones the planner will at the same time consider the type of action which is best suited to each zone and the targets which he will attack. When he has decided this he will be able to consider how many agents of various types he will need. These types can, roughly speaking, be divided into organizers, operators, technicians and guerilla fighters. As the whole question of selecting Agents is to be dealt with in another discussion I will not enter into it now.

At the same time the planner will be considering the equipment which he will require and the means of providing it and transporting it where necessary.

6. **PUTTING THE PLAN INTO ACTION**.

   The planner has now developed his plan, decided on his methods, subdivided his area, found what men and materials he requires. In the next discussion we hope to deal with the selection and training of those men, their despatch and their operation in the field.

SECURITY

1. **GENERAL**.

   Security means protecting yourself, as far as you can, against the attentions of the enemy. THEREFORE THE ONLY ANGLE FROM WHICH TO ATTACK A SECURITY PROBLEM IS TO PUT YOURSELF IN THE OTHER MAN'S SHOES.

   In this country, the main problems are to ensure that -

   a). Enemy agents do not penetrate the organization, or that details of our establishments do not leak out to them.

   b). The identity of our agents and operators, while they are in training, is not disclosed to the enemy.

   c). On the other side our agents should be in a position to protect themselves against the very powerful police organizations existing.

2. <u>RECRUITING STAGE</u>.

   Steps must be taken to ensure the bona-fides of every potential student by -

   a). Checking up his career with M.I.5 and other sources.

   b). A close study of the man's own character and story, namely that any person recruited in this organization should undergo a thorough interrogation covering his whole life which should be carried out by two officers, one of whom should be a senior officer of the Country Section.

   c). Recruits should be given Christian names at the outset by which alone they should be referred to throughout training or operations.

   d). In no circumstances should personnel of S.O.2 have two names. If they wish to use a false name they must do so consistently and in all circumstances. Having one name in the office and another for students is bound to lead to a show-down at some time.

3. <u>TRAINING STAGE</u>.

   All students should be given a talk on security by the Country Section before they start their training.

   The best kind of cover is to be consistent. To be inconsistent is to attract attention.

A. There are only two roles which the students can play, either on an all mufti or an all-uniform basis. Since the uniform basis has been chosen, it is highly desirable that everybody connected with them should be in uniform. To see a civilian apparently in charge of soldiers is to court surprise and enquiry.

B. Students must, therefore, represent British soldiers to possible onlookers. Points of importance -

   1. Their behaviour must approximate to that of soldiers, e.g. saluting officers, standing to attention, respect for N.C.O's, etc.

   2. Their dress must approximate to that of British soldiers, e.g. not wear collars and ties with battle-dress when in the ranks, brown boots, foreign greatcoats, foreign bolts, absence of gaiters, etc.

   3. It is better that they should have some kind of cap badge, e.g. A.M.P.C.

C. Cover. In the case of both staffs and students it is better to have a good story and stick to it, e.g. "Army Training". Avoid appearance of excessive discretion. To give the impression that one cannot give any account of oneself because it is too hush-hush will immediately arouse curiosity and speculation. Have a short simple story without embroidery.

4.   SECURITY OF THE AGENT IN THE FIELD.

There are three fundamental principles of all good security -

      a). Have a good story and stick to it.
      b). Avoid being conspicuous.
      c). Be simple and do not try to be clever.

A.   The only real safeguard for the Agent is to work up his cover at the earliest possible moment, to secure all necessary papers for it and to live completely the life he is ostensibly leading. He must never be out of character.

B.   In order to remain inconspicuous, it is essential to realize that all good C.E. work is founded on two main principles -

      1. Control, or Prevention, and

      2. Detection.

The main controls are -

      1. Control of Identity.
      2.   "   " Movement.
      3.   "   " Action.
      4.   "   " Communication.
      5.   "   " Press.

Thus an agent with no papers, or bad ones, may fall a victim to the enemy's control of identity; an agent who attempts to cross openly from the Occupied to the Unoccupied zone in France will be caught by the Control of movement; an Agent who circulates at night during curfew hours risks being spotted by the enemy's control of action, etc.

Detection is based on reports from informants of all kinds, 'agents provacateurs', investigations, observations, searches, etc.

The only way in which to defeat these systems is by conforming as strictly as possible to all the regulations, by remaining completely inconspicuous, and by not challenging attention in any way.

With the whole of Europe to police, the German C.E. system must now be spread very thin so that the activities of their Gestapo personnel will, therefore, be largely directed to cases sifted out for them in the first instance by the police of the Country or by paid or unwitting informants. Here are some of the ways in which conspicuous behaviours will arouse the attention of those around an Agent.-

    1.   May spend more money than he would be able to do in his chosen character.

    2.   The landlady may notice that the Agent is receiving an abnormal amount of correspondence; so may the postman, e.g. Mrs. Jordan at Dundee in the Nazi spy conspiracy.

    3.   He may be seen with people whom he would not contact in his role.

    4.   Various small points, e.g. table manners should conform to those of the country - his hands should

not be white if he is supposed to do manual work - he must not smoke expensive brands of cigarettes if he is receiving a modest salary - he must not order the wrong drinks etc. Remember, it is the small actions which often start suspicions.

C. <u>The third principle - Simplicity</u>. The Agent should be quite simple and straightforward in his actions and should not try to put the police off the scent once suspicion is aroused by various subterfuges of the Phillips Oppenheim variety, such as futile disguise, devious and inexplicable movements when followed, 'smart Alec' tactics in dealing with the police, etc. The safest procedure is to carry on quietly with one's ostensible life. The only other alternative is to disappear completely and quickly, if real danger threatens

The Agent's greatest weapon for disarming suspicion is to have built up in the first place confidence amongst the people with whom he mixes - a good reputation coupled with complete consistency of movement.

<u>Interrogation</u>. The Agent's main chance of withstanding successfully police interrogation is by strict observance of these three principles. It is essential that his cover (i.e. his ostensible life) should be so much a part of himself that he can think in the terms of that role and answer questions

without hesitation. The various contacts he has made will bear out the truth of his story when the police check up on him with them, as they doubtless will. The one thing the police will always seek to establish is inconsistency of conduct and will then proceed to fasten it upon the Agent and to exploit all the possibilities.

CONTROL OF A CIVIL POPULATION

1. INTRODUCTION.

    a). The civil population of a country under the control of an alien force is controlled by that force in order to neutralize and eliminate any elements hostile to the controlling regime.

    b). Study of the methods of control applied to the enemy in occupied territory will thus enable us to void being caught out by existing regulations and to find loopholes which give scope for agent's activities.

2. METHODS OF CONTROL.

    The enemy must try to detect subversive elements or failing that, to prevent agents being effective. This can only be achieved by applying the following measures to the whole population -

    a). Control Measures. By introducing regulations and controls the enemy seeks to make more difficult the introduction of agents into a country, to make conspicuous such agents as have been successfully planted or at least to neutralize their activities.

b). <u>Detective Measures</u>. As a supplement to the above controls, detective measures are applied by the authorities so that any agent becoming conspicuous owing to a contravention or evasion of regulations is immediately brought to their notice.

3. <u>CONTROL MEASURES</u>.

Control regulations will be introduced to a greater or lesser degree but will fall within the scope of the following categories -

a). <u>Control of Identity</u>. All persons compelled to have identity cards issued by the authorities who are, thus in a position to register and classify the population. The card may carry such details as name, address, description, signature, the stamp of the issuing authority, all of which enable the authorities to check up on the holder.

Additional documents for special purposes may also be in force. For example, passport, demobilization papers.

b). <u>Control of Movement</u>. The country is divided into areas which are designed to prevent persons moving outside their areas of normal residence unless they apply to the authorities for a permit. Such permits are not granted before the applicant's reason for travelling has been vetted.

Examples -

1. Prohibited or defence zones in areas of special military interest.

2. Protected places - munition factories, ports, aerodromes.

3. Special restrictions on the movement of certain classes. For example, aliens, Jews.

The agent must, therefore, either comply with the regulations giving a good reason for travelling, or risk evading the controls.

c). Control of Action. Prohibitions of various kinds tending to limit the scope for subversive activity (for example the carrying of arms, taking photographs, listening to the B.B.C., keeping pigeons, petrol rationing and the curfew,) or local regulations designed to make a stranger conspicuous, for example, special traffic controls.

d). Control of Communication. Limitation of the use of the telephone, censorship of cables, telegrams and letters, prohibition of wireless transmitters and receivers of a certain type.

e). Control of Publications. The press, advertisements, films, books, etc.

subject to censorship to prevent the transmission of information or the spreading of subversive propaganda.

4. DETECTIVE MEASURES.

The police and C.E. authorities try to locate persons contravening evading regulations by the following means -

a.). Informant Services. Information about contraventions or inconsistent behaviour obtained from persons used wittingly or unwittingly as informants, for example, landladies, barbers, waitresses.

b). Security Lists. The maintenance of records of persons who are potentially dangerous (based on the classification of the population cf. control of identity) whether resident in the country or known elsewhere. For example, communists, freemasons, leading democratic personalities, etc.

c). Snap Controls. Checks instituted at random on roads, railways, cafes, streets, Identity cards, travelling permits inspected, searches for arms, black market produce. House searches.

d). Penetration of subversive organizations by planting police agents, etc.

e). By use of Agents Provocateurs provoking subversive activity to enable police to locate dangerous elements.

      f).   Surveillance.

      g).   Censorship - postal, telephone, etc.

      h).   Radio interception and D/F.

5.   <u>CONCLUSION</u>.

The complete enforcement of the above controls will rarely be achieved as all C.E. organizations and measures must be tempered and fit with the priority requirements and capacity of the National Policy (e.g. economic life of the country, production, shortage of police personnel, employment of foreign seamen in U.K. or foreign munition workers in Germany.)

We must, therefore, study the efficiency and extent of existing controls, and the possible loop-holes which remain.

<u>SECURITY FOR W/T OPERATORS</u>

1.   <u>INTRODUCTION</u>.

Lecture deals with special aspects of security for W/T Operators apart from general principles laid down in "Individual Security".

2.   <u>CHOOSING OF PREMISES FOR WORKING SET</u>.

      a).   <u>Choice depends on</u>:

            i.   Security considerations.
            ii.   Technical considerations.
            iii. Combination of i. and ii. and district.

      b).   <u>Security</u>.

Safer to have number of sets dispersed over wide area with owners or occupants of premises recruited (see further below).

c). <u>Technical</u>.

Avoid steel-framed buildings. Key clicks easily audible in next room or if radio receiver working off same circuit. Consider aerial camouflage.

d). <u>District</u>.

   i. Thinly populated country districts, possibility for isolated buildings, e.g. farms, etc.
   ii. Towns - private house or place of occupation.

e). In case of d). ii. above, consider following factors:

   i. <u>Accessibility</u>.

      Operator must be able to get to and from premises without arousing suspicion of neighbours or passers by.

   ii. <u>Cover</u>.

      Must have "genuine" reason for frequent visits (e.g. doctor). Use existing household.

   iii. <u>Facilities, defensive</u>.

      For concealing self and set. For escape (exits). Vulnerability to surveillance.

        iv.   <u>Control of access</u>.

                 Limit to number and type of people with possible access to premises.

3.   <u>GENERAL SECURITY PRECAUTIONS</u>.

     To be taken in any premises including place of residence.

     a).   Precautions against search during absence - tidiness, leaf in keyhole, hair, etc.

     b).   Minimum incriminating material, coded writings destroyed, etc. N.B. traces on blotting paper and writing blocks.

     c).   Hiding places prepared, particularly for set.

          i.   Inside house - advantages and disadvantages.
          ii. Outside house - advantages and disadvantages.

     d).   Preparation for destroying incriminating material.

     e).   Where possible room with 2 doors and light switch near while operating.

     f).   Guard while operating, e.g. possibility of hall porter.

     g).   All clear and danger signals, visual and/or oral.

     h).   Check on surveillance of premises, or when entering or leaving.

i). Alternative premises in case of emergency.

   j). No casual visitors at premises - only possible ones are cut-outs.

4. CUT-OUTS.

   a). Definition.

   Intermediary. Link between two agents. May only carry messages, knowing nothing about organization, or act as liaison officer. Should undertake no subversive activity.

   b). Reason for Employment.

   i. Dangerous for operator to be seen with organizer.
   ii. May not want another member to know him.
   iii. Barrier between himself and authorities, e.g. telegram, official inquiry, hiring flat.
   iv. Transfer of suspicion, delayed or prevented.

   c). Cover.

   Must be able to contact inconspicuously people of

5. SECURITY RULES FOR OPERATORS.

   a). Must never undertake any subversive activity. Danger of overenthusiasm.

   b). Must not attempt to find out more about organization than he is told, nor know one or two members.

c). Christian names only should be used. Numbering dangerous.

d). Never carry arms unless in situation for which no cover story. (e.g. working the set.)

e). Must report suspicion incident immediately, e.g. if followed.

f). Emergency measures, e.g. warning signals, hide-out, contacts to drop, how to re-establish contact.

GERMAN C.E. METHODS PART 1 (PREVENTIVE)   C.E. 9

1. <u>Introduction</u>. Experience show that the German C.E. authorities are highly organized and methodical; impossibility therefore of leaving anything to chance. Conversely, the very rigidity of the system gives Agent certain loopholes, provided he conforms to regulations and controls and has an idea of methods used by the police and a cast iron story.

    Important to study the German Preventive measures known to exist.

2. <u>Control</u>. Can be divided roughly into two categories:
    (a) Direct.
    (b) Indirect.

    a). Identity - Identity Cards in use or new issues, if any.
    Movement - Limitation of travel internal or external.
    Action - The usual German method of increasing severity of regulations; c.f.

                          Increased hours of curfew, death sentences for arms.

            Communications - Suppression of telephones - wireless receiving sets, etc.

            Publications -

   b). By institution of small local regulations, with the object of making strangers conspicuous, e.g. walking allowed on only one side of pavement, bicycling to or more abreast forbidden. Sudden unannounced changes in ration cards and other official documents.

3. <u>Use of Local Population</u>.

   a). By making civil officials (mayors, etc.) responsible for safety of V.P.s, thus local population compelled to guard, and is solely responsible.

   b). Getting local Nazis (Quislings) to pick out pro-British elements.

   c). Ordering a selected number of latter to report daily to the Gestapo, with identity cards.

   d). Criminals released from prison on order of the Germans, to serve with local Quislings or as police spies.

4. <u>Use of Local Police</u>. In most territories, especially where cooperation from the population is sought, Germans seek to employ the local police for:

        a). Routine duties - Public order, etc.
        b). Sieves for information.

This is achieved by a comb-out of the local police, undertaken very soon after invasion. The general method employed in this comb-out is:

(a) Attaching a German policeman to work with a local man, with the object of picking his brains and locating his habits, weaknesses, etc.
Reporting on the local man after a period of about two weeks.

(b) Compiling a report of the experience gained by the various German police giving details of all local police who are likely to cooperate, be reliable or those who will be influenced by terror, loss of job, etc.

(c) Thereafter, leaving local police to sort out from mass of detail anything having flavour of espionage or subversion, which will immediately be handed over to Germans for further investigation.

GERMAN C.E. METHODS PART II (DETECTIVE) C.E. 10

1. <u>Introduction</u>. Basis of all detective work in C.E. lies in the informant service, thus Germans endeavor to use the local population as far as possible, as -

        a). Informers, unconscious.
        b). Informers, deliberate, e.g. Quislings, Collaborationists.

        c). Agents Provocateurs,
            Police Spies.

2. <u>Detective Methods</u>. All or any of these methods are known or may be used by German C.E. authorities either by themselves or in conjunction with the local police.

        a). <u>Surveillance</u>.

            i. Watching premises, railway stations, public places.
           ii. Following suspects.
          iii. Controlling hotels, lodging houses, taxis, etc.
           iv. Watching relatives and friends of suspects.

        b). <u>Interception</u>.

            i. <u>Postal</u> - Censorship in special areas. Generally no censorship inland, letters picked at random, or against suspect addresses, names on blacklist. Letters to neutral countries.
           ii. <u>Telephone</u> - Records taken of all calls. Telephone used fully as means of locating subversive activities.
          iii. <u>Telegraph</u>.
           iv. <u>Wireless</u> - By D/F they locate set within radius of 20 miles. D/F station moved about the area, narrowing it down until it is possible to

hear the sound of the actual key; this is possible at about 100 yards.
    v.   <u>Burglary</u> - from couriers, official mail, etc.

c). <u>Provocation</u>.

    i.   Bogus offers of service, e.g. to supply boats to facilitate escapes.
    ii. Passing on false information, e.g. rendez-vous for escapees.
    iii. Provoking acts of sabotage and resistance to identify more active anti-German elements.

d). <u>Penetration</u>.

    i.   Direct offers of service.
    ii. Indirect enlistment.
    iii. Double Agents.

e). <u>Interrogation</u>.

    i.   Prisoners of war.
    ii. Capture of actual Agents.
    iii. Persons whose names have been mentioned in other interrogations.
    iv. Kidnapping in Neutral Countries.

f). <u>Searches</u>.

    i.   House searches. Illicit radio, propaganda.

ii. Body searches. After arrest, at line of demarcation, etc.

INFLUENCING PEOPLE

1. INTRODUCTION.

   It is necessary for the agent to consider methods of influencing people from whom he may require assistance, protection or a specific service; this is distinct from the recruiting and handling of members or prospective members for the organization itself.

2. OBJECTS OF INFLUENCE.

   The agent sets out deliberately to influence people with certain main objects in view, viz -

   a). To make friends and thus obtain information and protection or help as and when necessary, even including a minor case of passive assistance, i.e. to establish a good reputation with the world in general so that people will vouch for him in emergency or remove suspicion if it exists.

   b). To make a person do an isolated act on his behalf, viz. lend a cart or bicycle, obtain papers, passes, supply money or food.

   c). To prevent a person from taking action against you, e.g. informing against you.

3. **DEFINITION OF INFLUENCING.**

   Influencing usually means getting people to do things for you in return for something which they themselves want, and falls into three main categories -

   a). Mental Influence.
   b). Material Influence.
   c). Coercion or blackmail.

4. **METHODS OF APPROACH.**

   The main incentive which makes people do things is because they want to. This may range from a man doing something because he likes you, to a man giving a bandit his watch because he wants to save his own life. The agent's approach is thus based on deciding what the contact wants and selecting his technique accordingly. He must, therefore, consider the three main categories in detail.

   A). **MENTAL INFLUENCE.**

   i. Smile.
   ii. Give the other man a sense of importance.
   iii. Be generally interested in others.
   iv. Be a good listener and encourage others to talk about themselves.
   v. Talk in terms of the other man's interests.
   vi. Give a man a good reputation to live up to.
   vii. Give the other man the credit.
   viii. Let the other man save his face.
   ix. Practise the "yes" technique.

(It is a psychological fact that if you can get a man to say "yes" twice he will say it a third time as this entails a physical and mental reaction.)

N.B. Don't merely persuade, but try to gain respect at the same time.

B.) <u>MATERIAL INFLUENCE</u>.

The background of physical influence is money. Do not, however, use money by itself unless it is absolutely necessary. Rather use the things which money can provide, e.g.-

a). <u>Services</u>.
- i. Communications, i.e. letters and correspondence with relations in occupied territories.
- ii. Social introductions.
- iii. Business introductions and contracts - business tips.
- iv. Employment.
- v. Release from prison or concentration camp.
- vi. Escapes.
- vii. Paying fines for people.
- viii. Medical assistance.
- ix. Lodging and accommodation.

b). <u>Products</u>.

- i. Food - cigarettes - petrol, etc. (black market) bearing in mind, for instance, special foods for invalids,

>   children, etc.
>   ii. Medicine and drugs, e.g. insulin for diabetes and any form of medicine difficult to obtain.
>   iii. Clothes.
>   iv. Scent, flowers and drink.
>   v. Foreign stamps.

C). <u>COERCION AND BLACKMAIL</u>.

In using an effective informant service one can often unearth a large number of details concerning people's private lives or past which they would wish at all costs to avoid being disclosed. All this is material for blackmail. An example of the sort of people already in receipt of this information is money-lenders. Another method of exercising coercion or blackmail is playing on people's weaknesses, getting them involved and then using threats of exposure. This is the technique usually employed in the recruiting of traitors, e.g. Baillie Stewart.

We must also consider specific methods of bribery and corruption. They can achieved in two ways -

a). <u>Overt Bribery</u>. Proceed with caution. Does the race among which you are working admit bribery? Is the prospect's profession recognized as corrupt? Is he himself bribable?

b). <u>Covert Bribery</u>. (Can be done in many ways.)

        i. Putting a man in position to take your money, e.g. forgetting your wallet in his office.
        ii. Selling goods to him below their value.
        iii. Buying goods from him above their value.
        iv. Losing to him at cards.
        v. Losing bets to him.

If the above methods appear too crude, good results can be obtained from expensive presents to his wife, etc. If you are dealing with a pseudo-respectable scoundrel he will probably like to deceive himself into believing that he is not being bribed at all. Do not mention the matter openly. "Decencies are preserved." Covert bribes give him cover if his finances are investigated. Do not tax him direct with having accepted your bribes unless unavoidable. Tactful hints that service are expected in return should be enough.

In communities where bribery is common or in dealing with a blatant scoundrel, direct methods can be adopted.

5. <u>CONCLUSION</u>.

All the above methods, naturally, must only be used in relation to existing circumstances and possibilities. In Occupied Countries there is naturally a vast amount of patriotism which could be prevailed upon without resorting to the latter type of methods mentioned. Nevertheless, these measures must be borne in mind as being often of extreme use in emergency. Therefore, being in the position to blackmail somebody is a great asset even if you are never obliged to make use of such information. Lastly, it should be borne in mind that the German armed forces are by no means immune from attack by other than the ordinary methods.

COVER

1.   DEFINITION OF COVER.

   Cover is the guise you assume to make it possible for you to do subversive work. This also entails a past history which must be interrogation-proof.

2.   GENERAL.

   Cover must always come first. Your guise must not be imperilled. Subversive activity comes, and must come, second to cover. In the long run it pays.

>   Example: A man must not neglect his cover work, even to do some subversive act. An engineer in a factory must put his factory life first. Let him carry out his "cover" duties. Once these are done, he can turn to his real job, like a banker to baccarat.

3.   SELECTION OF COVER.

   a). When selecting your cover remember that your job is receiving and transmitting of messages by W/T and passing them on to Agents. You must undertake no other subversive activity whatsoever. You may be in danger for the following reasons:

- i. That you are personally conspicuous.
- ii. From enemy D/F.
- iii. By working the set.
- iv. The set may be found.
- v. From your contact with persons whose messages you send and receive.

b). Your safeguards are:

- i. When not engaged in subversive activity to be completely inconspicuous.
- ii. Avoid having intimate friends.
- iii. Be sufficiently busy to have an excuse for not talking to people, yet have sufficient free time to visit places where you can pick up information.

Choose a job which will allow you to keep your set in a place where it will be well concealed and where you can have long warning of search.

Your job should also allow you time to work the set without being conspicuous.

c). <u>Freedom</u>.

When selecting your cover, try to secure for yourself the maximum financial and social liberty, freedom of movement and leisure.

- i. <u>Finance</u>: A stoke or dock labourer is tied economically. A boiler house foreman or dock superintendent has greater economic freedom. Whatever calling you adopt, you must live on the average pay of the trade or profession adopted in a similar style.

If you are living as a factory inspector who normally earns £ 6 a week, live at that rate only. Do not become conspicuous by living at the rate of £ 10 a week. You will be caught out if you do.

ii. <u>Social</u>: It is better to adopt a cover which will enable you to mix with all classes than one which restricts the circle of your contacts, e.g. commercial traveller, insurance agent.

iii. <u>Leisure</u>: In selecting cover, aim at one that will naturally give you a fair amount of leisure for your "hobby", and one that does not leave you physically or mentally exhausted at the end of your day's work.

Example: Often a man or woman on night work (starting, say, at 10 p.m. and finishing at 6 a.m.) will, after five or six hours' sleep, be freer than those engaged by day, with the added advantage of being able to communicate with those working by day and night.

d). <u>WARNING</u>.

It seems obvious to say:

i. Don't pose as a corn merchant if you don't know the first thing about corn.

ii. Don't say you are a mechanic if you never have handled an engine or lathe.

iii. Don't pretend to be a fisherman if your hands are not scarred, gnarled and knotted by hooks, lines and fishing tackle.

Yet these are three actual examples which in this war have caught out a British, French and German agent respectively.

Note: There is a danger about jobs providing perfect cover, as the enemy is fully aware of the advantages which they offer, e.g. journalistic.

4. <u>ESSENTIAL DETAILS</u>.

a). <u>NAME</u>.

Always use your own name if possible. If it is necessary to adopt an alias, live your alias and see that it comes as second nature to you.

Examples:

i. If a page boy calls out your alias in a hotel, train yourself to answer promptly.

        ii.   Should a passerby hail you by your assumed name in a street, train yourself to answer promptly, and not to react at all should an old acquaintance hail you by your real name.

        iii. Train yourself to sign your assumed name automatically.

        Remember - use of another name entails adopting the whole past history of another person, including details of schools, addresses, parentage, etc. Only in this way is there any hope of getting through an interrogation.

b).   HISTORY.

        Know every detail backwards. Let as much of it as possible be true. You may have to go out of your way to enact part of it. Remember that within five minutes of landing at your destination, you may be questioned by a hostile official. On your ability to tell your story convincingly your liberty and still more the success of your mission, may depend.

        Example: If you have been landed in a particular district and your story is that you made a three day cross-country journey to reach that district, the Gestapo (or Ovra) MAY ask you:

                "You have just come from L....?"

"Where did you buy your cigarets?"

"What cafes did you frequent?"

"What were the names of the proprietors?"

"Were the proprietors married or single?"

"Describe the building opposite your lodging."

You should be prepared for the possibility of their checking your answers in some detail. It is advisable, so far as is possible, to stick to the truth.

Such a detailed interrogation is unlikely if you play your cards well, but it is better to be ready.

c). <u>DOCUMENTS</u>.

Identity cards, ration cards, passports, demobilization cards, exemption from military service, etc. Examine these carefully and immediately they are given to you. Assure yourself that they are in order. Dates, descriptions, etc. Don't take anyone else's word for it.

It may be useful to carry unnecessary documents such as faked letters from friends (ante-dated), old references from employers on business note paper, etc., etc.

Note: It is possible to move freely for some considerable time within a country without papers (cf. Thurston). Where the job is worth this risk, it may have to be done.

d). CLOTHES.

It is important to study every item of your dress. Clothes must be right for your district, or as near as possible. Avoid brand new clothes. Buy secondhand if there is not time to "wear" them in. Remove all labels from clothing, ties, collards, underclothes, socks, hats, braces, suspenders; constantly remove laundry marks from linen if you are travelling about. Check up the possibility of invisible laundry marks. Procure genuine labels if these are available. Boots and shoes: take out markings in lining, wear good old ones in preference to new. And walk boots and shoes "in" as quickly as possible. Let their condition tally with your story.

People travelling with luggage should also go through their luggage personal effects, such as toilet articles, to make sure that, in the event of their wishing to suppress a certain part of their past history, they are not given away by a label on clothing from a country which, according to their story, they have never visited.

e) BEHAVIOR.

Here are some ways in which conspicuous behavior will arouse the attention of those around an agent:

i.   The landlady may notice that the Agent is receiving an abnormal amount of correspondence; so may the postman.

ii. He may spend more money than he would be able to do in his chosen character.

iii. He may be seen with people whom he would not contact in his role.

iv. Table manners, which should conform to those of the country.

v.   Tastes, e.g. cigarets, baths, caviar, drinks.

vi. Ignorance of local conditions - e.g. ordering things no longer obtainable, prices, transport.

vii. Slang - study present state.

viii. Beware of your own mannerisms - cut them out.

The Agent's greatest weapon for disarming suspicion is to have built up in the first place confidence amongst the people with whom he mixes - a good reputation coupled with complete consistency of movement.

DEFENSIVE MEASURES

1. **FINAL ARRANGEMENTS BEFORE DEPARTURE**.

    Reference to lecture on "Cover" par. "Clothes".

    Before starting your journey, make certain that the following precautions have been taken -

    a). **Body**.

    1) Avoid excessive hair grease.
    2) Does hair-cut conform with local usage?
    3) Moustaches and beards - are they right or wrong?
    4) Hand and nails; do they conform to your cover?
    5) No finger stains.

    b). **Clothes**.

    1) Be systematic; undress and place all clothes and effects on one table; examine each item separately; place on second table after examination.

    2) See that there are no jarring labels on any item - e.g. shoes, ties, collars, shirts, underclothes, socks, suspenders, braces, buttons, etc. NOTE: Marks denoting particular tailor must be avoided, but marks showing foreign origin not necessarily fatal as much foreign haberdashery was imported into all countries before the war. Although jarring labels should be removed, some labels are necessary to avoid suspicion.

       3) Laundry marks, visible or invisible.

       4) Turn all pockets inside out and remove dust, particles of tobacco, etc. by brushing.

       5) Shoe repairs, e.g. rubber heels.

  c). <u>Effects</u>.

       1) Remove from pockets all tell-tale scraps, e.g. bus and train tickets, English cigarettes, cigarette case, letters, photographs, fountain pens, pencils, lighters, matches, money, pen-knives, wallets, etc.

       2) Put back in pockets all articles which are consistent with cover (empty pockets would be suspicious).

  d). <u>Final check</u>.

      Make sure that you have your maps, compass, torch, first aid kit, water-bottle, flask, food, old sack and spade properly sharpened for burying parachute. None of these things, except the compass, should appear to have been made in England. The compass should be thrown away as soon as possible.

2. <u>WEAPONS</u>.

  a). Should an organizer or agent pack a revolver or automatic? Generally speaking: NO. For the following reasons -

       1) In most countries it is a <u>crime</u> to do so without a permit.

2) If one is stopped in the street or rounded up in a police raid, the mere fact of carrying a gun excites suspicion.

3) Nine times out of ten one can bluff one's way out of a serious situation without a gun.

4) Once a trigger is pressed the fat is in the fire.

   Exception: Carry a gun in a raid, but don't use it unless absolutely necessary. If you feel you must have the moral comfort of a gun, pack it away from H.Q., e.g. bury it against a raid.

b). It is better to use such weapons as -

1) Loaded sticks.
2) Rubber truncheons.
3) Knuckle dusters.
4) Knives.
5) Slip-knot scarves.

c). The following are available. Particulars will be supplied by the Country Section if they are to be part of your equipment -

1) <u>K.O. drops</u>. To be dropped in a drink and produces 8 hours unconsciousness. Takes 30 minutes to act.

2) <u>Benzedrine capsule</u>. A temporary stimulant. Effective for 5/6 hours.

3) <u>"L" Tablets</u>. Produce instantaneous death. To be taken in extreme emergency.

3. ENEMY POLICE ACTION.

   a).   If you suspect you are being followed:

   1) Give no sign of being conscious that you are followed.

   2) Test your suspicion by:
      Change of direction, Stopping to look at shop-windows, Entering large store and leaving by another entrance, Walking up a deserted street or two, Boarding tram, bus or taxi. In general, try to shake off shadower without appearing to do so. Have cover reason for taking all above precautions, e.g. before boarding bus, etc., look at your watch, make it appear from "unconscious" gesture of annoyance that you are late for something. Then jump on to bus.

   3) Do not go back to your H.Q. until you are sure that you have shaken off shadower.

   4) Do not panic. Being shadowed means that the enemy is not yet ready for action. You have time to make a plan.

   5) Do not go near any of your Agents.

   6) If possible try to have one of your own agents following you until he picks up your shadower and identifies him.

   7) Put into force all pre-arranged danger signals forthwith.

> NOTE: Make sure that you are really being followed, and that it is not merely a case of nerves or persecution mania, both of which are liable to occur.

b). <u>Arrest of an Agent</u>.

1) Warn those with whom he may be in contact. Stop all visiting of your people to arrested man's house or to his family's.

2) Endeavor through cut-outs or indirectly (through agent's lawyers) to discover reason for arrest. It may be for some small infringement of a petty regulation and have nothing to do with our work.

3) Through cut-outs look after man's family - for whatever reason he is imprisoned. This is a legitimate item of expenditure.

4) The enemy often makes mass arrests without any definite evidence. Therefore being stopped by police and asked for documents, or even arrest, is no evidence of discovery. Agent should keep his head and stick to his story.

5) Ascertain how much he knew.

c). <u>Emergency Measures</u>.

Have detailed plan ready in advance which can be put into instant execution. Keep this up-to-date with the growth of the organization.

1) Have alternative H.Q. prepared.

2) Pre-arrange cover if possible for key men to disappear.

3) Members of the organization personally known to the arrested agent had better disappear.

4) Arrange for disposal of stores, equipment and papers if they are threatened.

5) Plan new methods of internal communication, e.g. rendezvous, boites-aux-lettres, cut-outs, etc.

d). <u>If you are arrested and faced with interrogation</u>.

Cross-examination methods combine kindness, threats and cruelty. Your interrogators will try to sum you up psychologically. Do the same to them. You may decide to counter.

Kindness: With willing but uninformation loquacity.

Threats: With the determined assertion that nothing can be gained by threatening an innocent person, who is only too anxious to help.

Cruelty: See under.

If you succeed in keeping "mum", new methods will be tried such as -

1) <u>Microphones</u>. If your best friend is put into your cell, say nothing. Cell is certain to be wired to a microphone.

2) <u>Stool pigeon</u>. If you are allowed to mix with other detained persons in a common room, say nothing. Stool pigeons and wiring will be around you.

3) <u>Signed confessions</u>. If you are confronted by the signed confession of one of your agents, it is probably either a forgery or obtained either by trick or duress. Arrange in advance with your agents that if they are made to sign such a document under duress, they will use a special form of signature. If the signature is not "special" you will know by what means it was obtained and can point out that any document signed under threats is ipso facto valueless and untruthful.

4) <u>Drugs</u>. These are over-rated and need a skilled medical man to administer proficiently, and the results, from a cross-examiner's point of view, are scarcely ever satisfactory.

The Gestapo are known, however, to be using ether and questioning a victim immediately before he becomes unconscious. Counter attack to ether is COUNTING - from 1 to 100, and over again. When the question is put to you, you will only answer "thirty seven, thirty eight, thirty nine..." which is not entirely helpful.

5) <u>Hypnotism</u>. You can not be hypnotized into doing something which, in your heart of hearts, you do not want to do. Indeed, if you do not want to be hypnotized, there are

very few people who could send you into a hypnotic trance. If faced with any statement which you are allowed to have made under hypnosis, you are almost completely safe in denying ever having made it.

4. <u>FINAL MEASURES</u>.

If you are really "up against it" and realize that the enemy proposes to go "grilling" you until they extract all possible information from you, there are only two courses of action open to you, in justice to the member of your organization - an attempt to escape or suicide by "L" tablet, or other means.

<u>FINAL ARRANGEMENTS</u>

(Throughout this lecture students will be asked to make suggestions about important points and articles which might be useful.)

1. <u>INTRODUCTION</u>.

Throughout the period prior to departure you should go over the details of your story and your papers to become thoroughly conversant with each.

Sometime before your departure make sure that you will take with you everything that you will require and that there will be nothing which will not accord with your cover story.

It is necessary to carry out checks in the following order:

2. <u>BODY</u>.

Examine your person carefully.

a) Does cut of hair, mustache, beard, conform to cover story?

b) Does condition of hands, nails, feet, conform to cover story? E.g. manual workers have horny hands; the feet of people who have walked long distances show it.

c) Remove nicotine stains from fingers.

3. <u>CLOTHES</u>.

   Check carefully:

   a). Cut and quality - do they conform to cover story?

   b). Are they reasonably worn in?

   c). Replace unsuitable labels and remove washing marks.

   <u>N.B</u>. Tailor's markings to be avoided but clothes of foreign origin not necessarily suspicious, much foreign clothing being imported before the war into territories now occupied.

   d). Footwear - particularly trade marks, etc. Do not overlook repairs, e.g. rubber heels, etc. Remember that shoes worn by people having walked long distances show it.

4. <u>EFFECTS</u>.

   Examine all effects which you wish to take

with you, e.g. watch, rings, note-case, brush, comb, fountain pen, pencil, cigarettes, cigarette case, cigars and cigar case, tobacco, tobacco pouch, letters, photographs, etc. Leave behind those which are of unsuitable manufacture or incongruous. It is conspicuous to have no effects, so replace them in so far as may be possible with suitable articles, e.g. those purchased in places where you are supposed to have been - souvenirs, letters bearing out your story, etc.

5. <u>HABITS</u>.

    Check your personal habits and get others to consider them. Are they conspicuous? Do they accord with your cover story? e.g. table manners, way of dressing, handwriting, etc.

6. <u>PAPERS</u>.

    These will be perfectly forged, if not genuine; you should make sure that the details are correct, E.g. age, description, place of birth, visas, etc.

7. <u>EQUIPMENT</u>.

    Check over what equipment you will require. Consider the following -

    a). Map - is it the sort that you might obtain or be able to use in the country to which you are going, or should it be destroyed once its immediate purpose has been served?

    b). Compass and torch - where were they made? Can you retain them, and for how long?

c). First-aid kit, water bottle, flask - is it wise to retain these after you have landed?

d). Spade and sacking. These must be thrown away immediately after the parachute equipment has been buried. Make sure that you have material such as sacking on which to place the earth when digging the hole.

e). Suit-case. This should be of common appearance for the country to which you are going. Outstanding color or pattern would be conspicuous.

f). Money. Make sure that it is the kind current, E.g. silver coins may have been withdrawn from circulation. Useful to have sufficiency of small change. Notes of large denominations may attract attention, and where can they best be changed?

g). Food. If this does not conform to the country of your destination it must be eaten before you leave the 'plane or left in the plane'.

h). Any necessary accessories, as may be required according to the projected operation or circumstances.

8. <u>WEAPONS</u>.

a). Do you need a gun? It is generally only helpful when you are engaged in activity for which there can be no cover story, e.g. landing by parachute. At other times it is likely to be an embarrassment. If you take one, decide what to do with it after

>   landing.
>
> b).   Other weapons are less suspicious or conspicuous but sometimes equally helpful, e.g. knuckle dusters, loaded sticks, truncheons, etc.

9. <u>DRUGS</u>.

    Decide whether you will require any of the following:

    a).   "L" Tablets.
    b).   K.O. Drops.
    c).   Benzedrine.
    d).   "E" Capsules.
    e).   Ptomaine Tablets, etc., etc.

10. <u>FINAL CHECK</u>.

    Draw up list of all points and requirements so as to be sure that nothing is forgotten at the moment of departure. Carry out second check a few hours before you leave.

    a).   Body re-check - do not have excessive hair grease.
    b).   Clothes re-check - brush out pockets, avoid small pieces of tobacco, etc. Be sure shoes are clean.
    c).   Effects re-check - search clothes carefully so that articles have not been overlooked, e.g. bus tickets, coins, matches, etc. Has anything slipped through the holes in the lining of pockets?
    d).   Papers, equipment, weapons, drugs - check all over again; be sure that nothing has been overlooked.

THE ARRIVAL: THE FIRST FORTY-EIGHT HOURS

1. **ARRIVAL**.

    You have arrived at the scene of your operations. You have come either openly or covertly. What are the first things to be done?

2. **LEGAL ENTRY**.

    If you are arriving openly by land, sea or air -

    a). Conform with all the formalities.
    b). Do not try to smuggle in suspicious things which could be sent separately at less risk.
    c). Do not attempt any subversive activity too soon.

    Your legal arrival is an important and useful part of your cover story -- do not prejudice it.

3. **COVERT ENTRY**.

    a). <u>By Parachute</u>.

    i. On landing, look up and count members of the party, if there is more than one, and all containers as they leave the aeroplane.

    <u>Note</u>: All should fall quite close together. Aircraft should then give O.K. signal, which must be arranged in advance. Good idea, is little (paper)-funnel with a small light in it, can be seen for a long distance upwards.

ii. Disengage parachute, wrap tightly and camouflage.

iii. Join Leader.

iv. Check that all personnel and containers are safe. Search for any persons or thing missing, but do not run risks in doing so. For this and other purposes a call signal is useful, e.g. whistle with note of a night-bird.

v. Leader will give instructions for disposal of parachutes, flying kit and containers. They are usually buried. Earth should be shovelled into burlap bags and superfluous earth either emptied in a brook or if there is no better way, just thrown into the air for best disposal. Have receiving party bury hole beforehand.

vi. Be careful about the disposal of sacks and spades, especially if you are in a party. Several left lying about together are suspicious. Drop them all into a river or bury them. See above.

vii. Release carrier pigeon at dawn, if all is well.

viii. If you arrive with a container you will find it difficult and suspicious to transport. The container is conspicuous, its contents difficult to explain away. Carry as much as you safely can. Bury or hide the rest in the container. You will have to arrange cover to have your materials fetched. Be sure not only that you can find the place where they are buried, but that you can direct somebody else thither. You might hide the spade in the vicinity for digging them up. Spades are conspicuous to carry about.

b). <u>By Aeroplane</u>.

While pinpoint landings are almost impossible in occupied France they are actually done in the Balkans and the Yugosl. guerrillas actually are in direct contact with the allies in this way.

c). <u>By Sea</u>.

    i.   Beware of patrols and booby traps.

    ii.   Get off beach as soon as possible with all your belongings.

    iii.   Hide, sink or destroy boat. Best procedure is to fill in with stones, take it out as far as possible and puncture and sink it. Beware that incoming tides to not wash away your belongings on the beach. This has happened.

    iv.   Check your position on

                    arrival. You may be some miles from where you expected to be, but you will be able to find your exact position with map and compass. Ascertain it as soon as possible.

        v.   Go as far as possible inland under cover of night.

                NOTE: You will have to pass first a forbidden zone and then a restricted area.

<u>STAY OFF ROADS</u>.

  d). <u>By Land</u>.

        i.   Survey frontier and note frontier posts and landmarks. Learn details en route. If you have a guide he will help you. Do not walk back over the frontier again by mistake. This is easier than you think.

        ii.   Get at least 15 miles inside the frontier or beyond the nearest town, whichever is further. Then turn round and walk back. You will thus seem to be coming from the interior.

  <u>Distinguish here</u>: Between <u>Overt</u> and <u>Covert</u> Entry.

  <u>Overt Entry</u>: Don't try to smuggle cigarettes, etc.

4. <u>FIRST STEPS</u>.

    i.   If you move at night you are not

likely to be stopped as there are few officials about, but if you are stopped you will be questioned and will require a good cover-story. If you move by day you are more likely to be stopped as a matter of routine but less likely to be questioned as you will be merely one of many. Avoid market days when populations are controlled and searched on account of black market transactions. If you are in a prohibited area or one where any movement is suspicious, you must move at night. Get as much information as possible about this before you leave.

ii. Avoid officials until you feel confident you will have no difficulty in dealing with them, e.g., until you have a plausible explanation as to where you have come from; but if you cannot avoid them, it may be wise to approach them boldly with a question. This allays suspicion.

iii. On your way, from persons you meet, pick up all the information you can without their realizing. If necessary alter your plans. Be ready to improvise.

iv. If you lie up at any time, one man should remain on guard, if you are not alone.

v. Do not move in groups of more than two at most. Meet only at places where you have cover for meeting. Remember the principles of contacting.

5. **RECEPTION BY FRIENDS**.

    On arrival you may be met by friends. They will give you all information and assistance and probably put you up for a few days. If you are not actually met you may have been given the address of some person who will help. You will have to find him.

    Your information may not be up to date. Friends may have been arrested and re-placed by agents provocateurs. They may be watched or have been suborned. Have a description of the person and arrange a password. Do not reveal yourself to any member of his household, nor even to him, immediately.

    Have a cover story for your visit. This story may be useful to your helper also to account for your presence. Do not arrive at an unusual time. Choose one when you can avoid his friends, children or neigh-bours. Innocent gossip may betray you. Your meeting must be in accordance with principles of contacting.

    An early contact with friend(s) can be an important part of your cover story, make as much of it as you can. WARNING: Beware of contacting relations or intimate friends unless they are necessary to your mission, and you are certain of their dis-cretion. The temptation to tell of your return is sometimes overwhelming.

6. **WHERE TO STAY**.

    If you have no friends you will have to decide where to stay. You have the fol-lowing choice:

a). <u>Hotel</u>.

　　Advantages: Easy to find, moving populations, telephone.

　　Disadvantages: Usually watched. Staff must be loyal to police in their own interests, concierges, etc. Nearly always suborned. Sometimes rooms wired with microphones.

b). <u>Boarding House</u>.

　　Advantage: Less watched than hotels.

　　Disadvantages: Curious boarders. Difficulty of late return at night. Gossip.

c). <u>Lodging</u>.

　　Advantage: Far less surveillance.

　　Disadvantages: Not easy to find, not central, probably no telephone facilities. Gossip and spying from landlady.

d). <u>Paying Guest</u>.

　　Provided that you have a really good cover story which will account for your movements, etc. This can be good. An unsuspected and unsuspecting family will be excellent cover.

　　<u>Note</u>: It is often wise to stay in several towns and places before starting work, you will then have a "past".

       e). <u>Flop Houses</u>.

           Advantage: Good idea to stay overnight.

           Disadvantage: Prostitutes and owners have to "play ball" with police.

7. <u>FIRST FEW DAYS</u>.

    a). Lie low during the first few days and undertake no activity.
    b). Learn all about the place and the people.
    c). Build up your cover.
    d). Look about you for suitable informants.
    e). Start searching for permanent residence, suitable HQ, storehouses for material, etc.
    f). Start obtaining examples of documents, identity papers, etc.

<u>ARRIVAL AND FIRST DAYS</u>

1. <u>ARRIVAL</u>.

    An agent may proceed to his country of destination either openly or covertly.

    a). <u>Openly</u>.

        Important part of cover story so don't prejudice it. Perfect papers and reason for journey are implied. Comply with regulations (immigration control, customs, etc.) and do not attempt to take compromising material through these controls. It can more safely be supplied later.

b). <u>Covertly</u>.

    i.   By parachute. Carry out drill as instructed.
    ii.   By sea. Check position. Methods of concealing or destroying boat. Beware of mines, booby traps, patrols. Move inland by night (prohibited area).
    iii. By land. Survey frontier beforehand. Memorize landmarks. N.B. danger of recrossing frontier. Discretion in using services of guide (agent provocateur). Proceed at least 20 kilometers beyond the frontier or to a point beyond the nearest town and then turn back into the town thus appearing to come from the interior.
<u>Note</u>: Controls for coastal and frontier areas may be some distance into the interior - not necessarily on the coastline or at the frontier.

2. <u>FIRST STEPS</u>.

a). <u>Movement</u>.

May need to cover considerable distances. In general it is better to move by day except in prohibited areas. If stopped by day it is a routine matter; by night it may lead to serious interrogation.

Tips:

    i.   Don't move in groups of more than two.

- ii. At night all cars, motorcycles and bicycles are potentially dangerous (facilities granted to Germans and collaborators only). Therefore always hide in this emergency.
- iii. Times to move: beware of curfew and dusk hours; best to move at rush hours; special care on market days.
- iv. Avoid long-distance trains and try to break journey. N.B. If cover permits, travel First Class. Small country stations inadvisable as stranger is conspicuous. Care at large termini and junction stations for regular controls. Possibility of alternative exits.
- v. Do something definite during day, particularly when in towns. Don't hang about.
- vi. Lie low if Allied Plane shot down or any disturbance (escape, shooting, parachute scare, etc.).
- vii. Where possible avoid carrying parcels or suitcases except when they are demanded by your cover.

b). <u>Cover Story</u>.

- i. Explanation of present and recent movements and activities essential (especially while moving from landing point to nearest town).
- ii. Necessary to have arrived from somewhere else. Therefore

                iii.   check the details, e.g. train times, contained in your explanation as soon as possible.

                iii.   Clothes, state of shoes, etc. must not belie the cover story.

    c). <u>Facing Officials</u>.

        In early days avoid contact with officials whenever possible. If meeting face to face with an official, better to approach with a question than to "run away".

    d). <u>Obtaining Information</u>.

        Seek information unobtrusively all the time re regulations and conditions. Watch particularly for signs of an unusual situation, e.g. state of alert, special controls in force due to disturbances, etc.

3. <u>RECEPTION BY FRIENDS</u>.

    a). If met by reception committee, many initial difficulties disappear.

    b). If not met but supplied with a contact address, make this contact as soon as possible (will give cover, information, possible job, etc.)

        i.   When contacting a stranger, do not rely only on name and address, use a description and if possible a password.

        ii.   Check to see that information

        re contact is up-to-date, e.g. still at address, not under surveillance or "blown".

   iii.   Prepare cover-story for visit. If possible make it part of your general cover. If you don't know the man, do not pretend you do.

   iv.   Choose time for contact carefully - beware of revealing yourself to other members of family, servants or children, except within your cover.

   v.   Always beware of contacting your relatives, particularly your parents, except when they are likely recruits.

   vi.   Don't take messages in any circumstances.

4. <u>WHERE TO STAY</u>.

Agent may have to take temporary residence. Choice of place is primarily determined by his cover. Following possibilities should be examined:

   a).   <u>Hotel</u>.

       Advantages: crowd, moving population, inconspicuous, gives freedom of movement. Disadvantages: usually watched, danger of informers. "International" hotel is not suitable for protracted stay but smaller hotel may be if cover permits.

b). <u>Pension</u>.

Less easily watched by the police. The disadvantages is lack of freedom of movement and curiosity, gossiping, etc. of boarders. If thoroughly reliable pension keeper is known, this could be used.

c). <u>Lodging</u>.

Difficult to find and stranger is conspicuous. Landlady may be curious and report to the police. If lodging is taken, try to win landlady's sympathies early by some plausible story.

d). <u>Paying Guest</u>.

Difficult to arrange for temporary residence but is possible for permanent purposes, provided the family is unsuspected. They may have to be in the know.

e). <u>Villa</u> (e.g. France or Belgium)

Good for permanent residence, independent, away from outsiders. Better not to live alone; best cover is woman.

f). <u>Charitable Institution</u>.

For seamen etc., but considerable danger of mass arrests.

g). <u>Brothels</u>.

    Not considered suitable but in real emergency. Danger of informers and police supervision.

h). <u>Prostitute with Flat</u>.

    Possible in emergency and certainly better than brothel. Dangerous type, as they can easily be bought and are police informers.

5. <u>EARLY DAYS</u>.

  a). Lie down and undertake no subversive activity.

  b). Build up your cover - job, recent history. Let your actions constantly create cover.

  c). Check up on documents and C.E. control.

  d). Look for permanent residence.

  e). Check landing grounds if given instructions on this point.

  f). Make arrangements for communication with W/T operator or other member of organization already on the spot.

  g). Begin survey of area.

      All the above will involve obtaining information by use of personal reconnaissance and the informant service that must be established from the outset.

ESCAPE PROCEDURES

1. INTRODUCTION.

   a). The place of this lecture in the "A" Material as a whole - its special relationship with the preceding lecture - that on arrival.

   b). Purpose of this lecture. Not that of prescribing any one technique or device as invariably successful; no two areas are alike and conditions are ever changing. We point out certain general principles and show how they have worked in a few specific instances. Agent must exercise individual judgment in applying the general principles to solving his personal problem.

   Also one other word before we get down to details. In some areas the underground may be equipped to direct and control the agent's escape (c.f. lecture on underground). Obviously to just such degree as it assumes this responsibility the agent's task is made easier; but for the purposes of this lecture, we shall assume the agent is acting largely on his own. Now to the problem.

   c). The Problem: Escape as distinct from departure. Escapee's cover blown or about to be blown - danger - the net about to be drawn. What is the answer?

2. HIDEOUT.

   (Known to no other member of local organization, or possibly in quarters of man whose sole connection with OSS is furnishing hideout). Possibly alternative hideouts.

3. <u>MONEY</u> - in several special caches unknown to other members of the organization.

   a). Small denominations in general befitting escapee's clothes - if feasible, some of it gold. Denominations retired must be replaced or at least remembered as no good. (Story Dutch 1000 Guilder note).

   b). Possible hiding places: hole beneath fence post; bannister; cemetery; greens of golf course (Germans in Westchester) in #10 tin wrapped in weather-proof covering usually used for weapons, the resulting package being buried (Far East). See lecture on Searches for further possibilities.

4. <u>CLOTHING AND EQUIPMENT</u> - including necessary documents also cached in advance.

   a). Vary with place and conditions, but <u>clothing</u> must be of type to be expected along route of escape - usually old, worn, and of a type sold in the country in question. Do not <u>look</u> foreign. "If you have new clothes, you have done something wrong".

   b). <u>Documents</u> - usually as a minimum: identification card, travel permit, and ration cards. Personal letters directed to identity indicated on card have been a great help on occasions. Sometimes requisite documents can be obtained in black market; sometimes seals can be obtained through bribery of one or more in office of local officialdom; documents are part of underground (c.f. News Digest, Jan. 18, 1944 for story on head of Danish Passport office and his deputy fleeing to Sweden - had aided great number in escaping.)

  c). Sometimes <u>means altering personal appearance</u>. (C.f. They Shall Not Have Me - Helion), <u>food</u> in form which will not spoil, <u>fuel</u> for motor (c.f. White and Spain).

5. <u>TRAVEL ACROSS COUNTRY</u>.

  a). General precautions and suggestions:

    1. Have your cover story well in hand and memorize your route of travel - do not carry maps.

    2. Carry as little baggage as possible.

    3. If at all possible secure aid of underground. Most escapes aided at one stage or another by underground. (See lecture on Underground) But beware of "false undergrounds."

    4. In making contacts helpful to have chocolates or cigarettes to give away, but do not show too much at one time.

5. Make your inquiries among poorer classes but look out for bootblacks, cafe workers, beggars, prostitutes, and agents disguised as loafers. Some escapees feel that frequent questioning of the police is good technique, especially if one thinks he is being tailed - "The best defense against policeman is another policeman." One man who escaped suggests that "The safest way for anyone starting on a journey to find out about German points of control would be to ask an elderly woman of the lower classes or a farmer in the most casual way, 'where the Boches were controlling.'"

6. In small towns in Catholic areas, parish priests most apt to help. (Many priests work with underground.)

7. If stopped and questioned, an excuse which frequently works is that you are looking for food and have been told "a farmer over there has food for sale."

8. If circumstances force you to stand around and wait at some place, the pretense of work (as in ruins) may give cover.

9. If out after curfew and stopped by a sentinel a pretense of drunkenness or clandestine love affair may help. (Story to illustrate.)

10. In many cities there are rough areas relatively free from police. For example, two boys escaping from Holland to Spain made one of their stops in a hotel in the Negro quarter of Bordeaux. "As many German soldiers had been killed in this quarter. The Germans prevented them (i.e., the Germans) from entering it. It, therefore, was the safest hiding place for refugees as no Germans are seen in the streets except during infrequent police raids."

11. Street Controls frequently omit apartment houses. On the other hand, if the control is a block control, keep moving in the street.

b). Some of the means of travel and comment upon them.

1. Travel by foot makes it relatively easy to avoid many controls - taking remote paths, crossing fields, etc.

2. Travel by bicycle has proved relatively safe.

3. Buses are less subject to check than trains but more crowded - passengers more subject to prying conversation and questions. General suggestions for those using either bus or train are:

a. Don't take bus or train from terminal; terminals generally watched carefully. Some escapees suggest that terminals should be avoided entirely - even if it means leaving train some fifteen miles from terminal, walking past it, and boarding another train some fifteen miles beyond.

b. Take locals and travel by roundabout routes when "Things are hot." Leave train or bus before it reaches border. One escapee recommends round-trip ticket to near border. Almost all borders and coasts under German control are not restricted or forbidden zones - special German army permits needed.

c. If one purchases single tickets for short distances, he can at any time pretend he is making only a local trip - either to depopulate a given town or to get work.

d. Avoid any but necessary conversation. Appear sleepy or use the tobacco chewing trick.

- e. The pretense of sleep and showing whole mass of documents (passport, permits, ration cards, ticket, etc.) at inspector at once makes him careless in checking them.

- f. Avoid secluded corners in stations; the police give them special attention.

- g. Controls sometimes can be avoided by entering or leaving station through a lavatory with exit into adjoining hotel (St. Charles R.R. Station in Marseilles - similarly in Toulouse).

- h. In some stations controls are enforced only during the period shortly after train arrival. Going to restaurant or library before leaving station frequently means to avoid controls entirely.

- i. Sometimes controls can be escaped by riding the steps of the train.

6. CROSSING BORDERS.

Usually borders are crossed with the aid of passeurs, sometimes with connivance of border guards, e.g., between Norway and Sweden. Suggestions are:

- a). Be prepared to pay high price for guide services and follow instructions carefully.
- b). Take time for reconnaissance - Note condition before making attempt if at all feasible. Sometimes well to work for local customs official for time or hold job requiring occasional crossings of border. (Woodworker, charcoal burner, etc. good on Franco-Spanish border).

- c). May be possible to cross after patrol has passed or during change of guard.

- d). Sometimes possible to get local man to distract guard's attention while refugee crosses and escapes.

- e). If passage by boat is involved, be prepared to change plans if the weather makes it necessary. Also, if possible, reconnoiter shore and learn tides, currents, and winds in advance.

7. <u>CONCLUSION</u>.

In conclusion let me point out that the details of escape usually <u>must be developed from within a country</u>. Only there is the best and latest information available on travel conditions, controls, and so on. As we said in the beginning, this lecture does not pretend to offer sure-fire techniques of escape. By example and suggestion it presents the problem and indicates possible solutions under specific circumstances. It is for the individual agent to choose and develop the possibilities which offer the best chance of success in his particular situation. At best this lecture can only help him.

COMMUNICATIONS - PERSONAL MEETINGS

1. GENERAL.

   Advantages.

   a). More information can be passed.
   b). Correct emphasis on information can be made.
   c). Immediate replies can be obtained.

   Disadvantages.

   a). Meeting may be suspicious.
   b). Suspicion may be passed on to hitherto unsuspected contact.
   c). Meeting may be a trap.

2. METHODS.

   Three types of personal meeting are discussed with a view to obviating disadvantages.

   a). BETWEEN ACQUAINTED AGENTS.

   1) Cover for meeting.
      - By chance. (First arrival mustn't show impatience.)

      - On purpose. (No confusion with chance meeting.)

   2) Choice of meeting-place.

      Bad: Railway stations, public meetings, hotels, brothels, queues (Informers), post-offices. (If you must meet at one of these, use danger signs.)

      <u>Better:</u>   Small hotels, restaurants, bars,

      <u>Still Better:</u>
                    Street, gardens (exc. France), Catholic churches, cemeteries, Turkish or Swimming-baths, museums, art-galleries, plages, parks, motor-cars, house hired or rented by third party, private houses and offices, "hangabout" places (c.f. woods near Oslo on Sunday); and dance-halls or places of amusement when not controlled (c.f. no dancing in Norway).

      NOTE: Always have cover for your presence at any meeting place.

3)   <u>Conversation</u>.

      Cover conversation. Arrange on arrival.

      Don't whisper except e.g. for dirty-joke or love-making.

Smile and laugh as much as possible (Permanent tip for agent.)

4) <u>Passing messages</u>.

No tricks, if it is possible to pass message unnoticed. (e.g. in W.C. etc.)

If tricks necessary, take equipment. (Use of identical equipment - swapping briefcases, matchboxes, newspapers, etc.)

Sign to indicate message passed.

5) <u>Security Precautions</u>:

- Innocent pre-visits to R.V. as cover.
- Security of pre-arrangements. (e.g. time and date given separately and not by same method such as telephone. Communist trick of ADDING one hour to time - therefore, police are always late.)
- Safety and danger signs (e.g. NO gloves - danger.)
- Alternative R.V. in case of accidents.
- <u>Be punctual</u> (Synchronise watches or pre-arrange that time be taken from neighbouring public clock.)
- Guard against microphones. (Turn on wireless, bath, gramophone.)
- Limit the time you are going to wait.

- Precautions against surveillance. (Meet in black-out where practical.)
- Change R.V.'s frequently.

b). <u>BETWEEN UNACQUAINTED AGENTS</u>.

    1) <u>Cover</u>.

        - Must be chance meeting.

    2) <u>Description</u>.

        - Proper description better than combination of signs. (Wearing red rose and drinking green drink.) Signs should be used only as check for description.

    3) <u>Passwords</u>.

        - Never conspicuous - i.e. dramatic, incongruous or compromising.

        - Therefore generally banale but marked, esp. in reply. (Use of pause, gesture, special but commonplace word.)

        - Should be short and spoken verbatim. (Easy to forget passwords in stress.)

        - Should be introduced naturally. (Some passwords are good 'openers', others have to introduced gradually.)

        - Reply at once.

(E.G. One agent used a badly chosen password to a M.V. staying at a certain hotel. V. looked astonished, and the agent later found that there were three men in the hotel having the same name.)

c). BETWEEN AGENT AND OUTSIDER.

Security precautions.

- If you're suspicious, don't go to R.V.

- Don't accept outsider's suggestions on time, place, date. Send cut-out to R.V. to bring him somewhere else.

- Intercept him yourself en route, and take him elsewhere.

- Send man to receive R.V. in advance.

- Have yourself followed.

- Extra search of yourself before meeting.

COMMUNICATION - INTERNAL

1. INTRODUCTION.

Organizers must be able to keep in constant touch with colleagues. It will be found that better work is obtained if all subordinates feel that means of communication are assured.

It must be remembered that in communicating an agent endangers himself and friends. Therefore greatest care is essential.

It is necessary to study the wide choice of methods available, adopting the most suitable method to the circumstances.

2. <u>POST</u>.

Much the simplest way for persons at a distance to communicate with one another, but slow, uncertain and fairly easily investigated by police. In occupied territories letters frequently censored, especially in and out of prohibited areas. Letters of suspected persons sure to be opened; thus consider the following measures:

a). Essential to use code (veiled language) or secret inks.
b). If suspected, do not use post for subversive correspondence.

c). Avoid putting precise address at letter-head, except where regulations demand it. If unavoidable, existing address (e.g. real hotel, innocent address, third party, Quisling) is preferable to an invented one.

d). Avoid watermarked paper and typewriters of non-standard makes or which have recognizable peculiarities. Only use your typewriter for secret work. Hide it when not in use.

e). Sign with christian name or nickname.

f). Avoid special classes of mail (air, registered, parcel post). Make use of post cards and business mail, particularly circulars.

g). Post at different post-boxes and at places away from place of residence. Remember sender's letters can be identified if sender is followed (e.g. newspaper thrown into post-box after letter.) Therefore take precautions against surveillance.

h). When sending instructions by post, allow for delay in delivery owing to war-time conditions.

i). Greater security is afforded by accommodation address, especially if it gives cover for the reception of numerous letters, where organizer has no such cover; conceals your address and identity from your correspondent. Letters must be sent as if no accommodator; no enclosures and passed on by hand or post according to purpose of accommodation address. Special outside markings useful to show that letter contains secret message. Change accommodation address frequently. Only use for two correspondents at most. Avoid poste restante- such mail is watched; only use when no other address available.

3. <u>BOITES-AUX-LETTERS</u>.

These are places where persons or their cut-outs can leave and collect messages. A useful means of communicating between principals who do not wish to meet. Very difficult for the police to trace one party by surveillance of the other.

They must afford cover for frequent visits, e.g. tobacconist, cafe, newspaper kiosk, etc.

Remember:

- a). Intermediary must be recruited but need not be informed of the nature of the organization.

- b). Correspondents should hand in letters clandestinely, or have cover for leaving them openly. Study ways of handing over messages, e.g. in newspaper, in food, with the change, etc.

- c). Decoy letter and/or alternative cover (e.g. black market) should be available in case of emergency.

- d). There should be danger signs in case of emergency.

- e). Clear box regularity.

4. "DEAD" BOITES-AUX-LETTRES.

Same as above, but without living agency, e.g. milestones, lavatories, etc. Here again principals can communicate without meeting.

NOTE:

- a). Three is no risk from intermediary.

- b). There would be no warning in the event of discovery, in which case the place would be watched by C.E.

        c). Have cover story for "finding" message, e.g. find by accident.

        d). Police might watch correspondence without principal knowing.

5. <u>COURIERS</u>.

    A courier supplies means of circumventing censorship by having messages carried by hand or verbally. Slow but surer than other methods. Message can be destroyed in emergency. On the other hand, if courier should be captured and interrogated, there may be imminent personal danger to several organizers. Precautions:

        a). Courier must have good cover and know route well.

        b). Message is best memorized, but if this is not possible it should be written on paper easy to destroy and/or conceal, e.g. silk paper, rice paper, cigarette paper. In such circumstances.

        c). Courier should not know contents or code.

        d). Courier should be instructed in the principles of body searches and snap control. Courier should not have direct contact with principals.

        e). Relay system of short-stage couriers connected through boites-aux-lettres has advantages. Each courier knows stage intimately and should not exceed his stage.

6. "DEAD" COURIERS.

   Messages can be concealed on vehicles or in luggage of unknown persons. Same conditions as in case of "dead" boites-aux-lettres.

7. TELEGRAMS.

   An easy and quick method of sending short messages between persons at a distance, but very conspicuous and certain to be censored, and of limited use only - to be used sparingly. Sender will almost certainly have to identify himself. Two permanent records of contents. Telegrams a good means of indicating the whereabouts of the sender, arranging rendezvous and answering questions.

   Precautions.

   a). Code with good cover must be used.

   b). Use cut-out to transmit wire as his own.

   c). Make sure recipient socially qualified to receive telegrams.

   d). Accommodation address (see Post).

8. TELEPHONE.

   Very quick method for those at a distance to communicate, but always assume that telephone conversations are censored. Conversation of suspected people often recorded automatically.

   Therefore use public call-boxes. Trunk callers may have to identify themselves -

serious limitations on trunk calls may exist.

Use of foreign languages generally forbidden but careful code must be used, and conventional conversation. Suspicious phraseology likely to be noted. Telephone can best be used for arranging rendezvous, answering questions, etc.

9. ADVERTISEMENTS.

A means of communicating reasonably quickly with a large number of people simultaneously. Could be used either as a sign or to conceal coded message. Impossible to identify recipients. Best used as general warning or action sign. Remember:

a). Identity of inserter must usually be established. Possible to use cut-out.

b). Wording must be normal as it is likely to be carefully scrutinized by C.E.

c). Good cover for inserting essential. Articles lost preferable to "for sale" or "goods required" notices.

d). Regular insertions must be made by regular advertisers.

e). Delay.

10. GENERAL PRECAUTIONS.

a). Messages can often be sent in two parts, either part alone being incomprehensible, e.g. the main message by courier, the key by post.

- b). There should always be at least two means of communication between important organizers working together, in case one is "blown".

- c). Important messages should be sent in duplicate.

- d). May be necessary to arrange for acknowledgment of message.

- e). Avoid numbering of messages.

COMMUNICATIONS - EXTERNAL

1. **GENERAL**.

   Communications between agent in field and base are vital and require careful organization. Several alternative routes are advisable in case one is "blown". The following are the principal methods:

   - a). <u>W/T</u>. Short, urgent messages and/or where immediate reply is required.

   - b). <u>Courier Service</u>. For long messages, urgent and/or less urgent; maps, material, etc.

   - c). <u>"Innocent" Letter</u>. Sent to address in neutral country. Short messages for use when agent is out of touch with members of his organization.

   Other possible methods are:

   - d). Telephone or telegram to neutral country.

   - e). Advertisement in press.

       f). Diplomatic bag.

       g). Carrier pigeons.

2. <u>W/T</u>.

This is the only method for rapid communication and for obtaining immediate reply. The more it is used the more likely it is to be detected. Operator is highly trained man with special cover. To protect him and/or his activity, the following precautions are required:

       a). He should not be used for other work.

       b). Other agents should not go to his residence or place of operation. It is even better if they do not contact him direct.

       c). A reserve means of communication with him must be maintained if he has to go into hiding.

       d). Only messages which cannot be sent conveniently by other routes should go by W/T.

       e). Messages must be between 150 and 400 letters.

3. <u>SECURITY DEVICES OF OPERATOR</u>.

       a). Set disguised as suitcase.

       b). Bury set on arrival.

       c). Aerial made of local wire, camouflaged.

- d). Must have cover which allows absence at irregular times. Routine job useless unless employer in organization.
- e). Should live with friends as key-taps audible. Residence must supply hiding place.
- f). Should constantly move set and/or aerial.
- g). Times and length of transmission restricted according to plan.

4. <u>COURIER SERVICE</u>.

    A series of linked relays by which messages and/or material are carried. There may be several "termini", each linked with a "branch service" or "main service". It is much slower than W/T, but usually safer. In general, couriers cover each stage.

    The following points should be noted:

    - a). Each courier should have special knowledge of his relay and good cover for travelling regularly.
    - b). Several couriers can be employed on each stage so that there is a frequent service. Often courier will make journey innocently.
    - c). Couriers contact with one another and principals should be through boites-aux-lettres where possible.

5. <u>DIFFICULTIES AT FRONTIER OR SEA</u>.

    Special problem arises at a crossing of

frontier or sea. The following methods are used:

a). Smugglers, attendants on trains, lorry drivers, etc. used as couriers.

b). Trains, cars, other vehicles, used as "dead" couriers.

c). Material floated down rivers or thrown over frontier line in containers, e.g. "turnips".

d). Drum taps and puffs of smoke, etc.

e). Neutral seamen as couriers.

f). Materials smuggled out to submarine, seaplanes, fishing boats, etc.

g). Material concealed on ships or aeroplanes, etc. by dock or aerodrome workers, etc.

6. LETTER SENT TO ADDRESS IN NEUTRAL COUNTRY.

Safe, but slow method. Letter can contain:

a). Long message in secret writing.

b). Short message in code.

c). Pre-arranged signal, e.g. warning, "all well", etc.

d). Agent's address when he is out of touch with the rest of his organization. This is the commonest use:

Precautions:

i. Addressee does not know purport of letter but believes himself intermediary between innocent person in occupied territory who wishes to write to friends in allied territory and cannot communicate direct. Much innocent correspondence, in fact, passes this way.

ii. Sender must take every precaution against censorship which is stricter for foreign letters than for internal, e.g. choice of address, danger of registered letters, air mail, length of letter, etc.

iii. Not too many letters must be sent to the same neutral address.

iv. It is more important to conceal identity of sender than recipient, who runs little danger.

7. TELEPHONE.

Telephone messages to and from neutral countries are sometimes permitted. Formal application must be made. Use of language restricted. Censorship certain.
Therefore good cover essential.

Use of code highly dangerous except between experts. In general only suitable for arranging rendezvous, giving address, etc.

8. TELEGRAMS.

   Same considerations apply as to telephones - see previous lecture. Telegrams are useful for indicating address of sender.

9. ADVERTISEMENT IN PRESS.

   Enemy papers available in this country within a week. First class cover required for frequent insertions. See previous lecture.

10. CARRIER PIGEONS.

    Rarely used except for notifying safe arrival. Special instruction essential.

    a). <u>Advantage</u>. Can take small maps, drawings, etc., quicker than any other methods, except W/T.

    b). <u>Limitations</u>:

       i. Difficulties of importing and hiding.

       ii. Difficulties of feeding and exercising.

       iii. Loss of homing instinct after a time.

       iv. Limitation to distance and direction which can be flown.

       v. Pigeons cannot 'home' in darkness or in fog.

APPROACHING AND RECRUITING AGENTS

1. NEVER be rushed in recruiting agents, no

matter how badly you need them. Even if you are sent to your home town no one can reasonably expect you to create an organization in a week.

2. On arrival, plan your recruiting campaign as follows -

      i) Survey of locality.
      ii) Consideration of types needed.
      iii) Survey of prospective agents.
      iv) Approach to prospective agents.

  i. <u>Survey of Locality</u>. Whilst establishing your cover, make as many normal contacts (to be used as unwitting informants) as are consistent with your cover. From them and any other source discover -

      a). State of local opinion - how far collaborationist, how far anti-axis.

      b). Industries, occupations and conditions of workers; relationship (i.e. rivalries) between classes -both among themselves and with the enemy.

      c). Information on influential local people and their subordinates. Which are reliable, which traitors. Their private affairs - financial - social, etc.
          Note: Lists of traitors are available in England. You should try to identify all those in your district.

ii. <u>Consideration of types needed</u>. You cannot possibly decide, until you have a knowledge of your locality and its inhabitants, what sort of agents you will want. This also depends on your specific task. The following types are likely to be needed.

    a). <u>Insiders</u>. If you are going to work in an industrial group, close associates or members of that group are essential.

    b). <u>Specialists</u>. For sabotage, engineers and mechanics, for propaganda, printers and journalists.

    c). <u>Cut-outs</u>. People who can contact two (perhaps socially unequal) persons without arousing suspicion: e.g. doctors, chimney sweeps, travelling salesmen, charity workers, officials, priests.

    d). <u>Boites aux lettres</u>. People in charge of places frequented by all classes - e.g. shops, kiosks, post offices, etc. which can be used for depositing and picking up written messages.

    e). <u>Accommodation Addresses</u>. E.g. respected business men who receive letters through the post regularly at their office.

    f). <u>Couriers</u>. People who can travel without exciting suspicion over a particular territory. E.g. commercial traveller, engine or lorry drivers, etc.

g). <u>Collectors of "imported" material</u>. People with good cover for being out late at night and of the smuggler type. E.g. poachers, frauders, gamekeepers.

h). <u>Storers of material</u>. People who have access to premises unlikely to be searched or too frequently visited, -e.g. farmers with barns, wine merchants with cellars, shopkeepers with stock-rooms.

i). Persons who can supply you with an H.Q. or a reserve H.Q.

j). <u>Women</u>. In some jobs women are better than men, e.g. making both men and women talk or as intermediaries.

iii. <u>Survey of Potential Agents</u>. It is from the ranks of your informant service that your first recruits are most likely to be drawn, and many of the remainder will be indicated to you by the same means.

Put quality first; a bad agent will jeopardise your organization. Therefore, you will probably only select a small proportion of the "possibles".
Remember -

a). Get to know all you can about

a potential agent before approaching him.

b). Get to know him and find out his weaknesses and interests - e.g. wine, women or stamp-collecting. Have you any sort of hold over him, should he go bad - e.g. his wife and children. (Note: Agents, however, loyal, who have relations in the power of the enemy, may be dangerous.)

c). It is better to recruit agents on the grounds of their political, religious, national or racial convictions than to buy them. (Note: political or religious creeds can be a drawback, e.g. Protestant in Catholic country, freemasonry.)

d). People who are relatively comfortably off and therefore have something to lose are sometimes poor material. Conversely, those who have already suffered heavily often make first-class agents, e.g. Poles.

e). Persons can sometimes be used unwittingly, e.g. acquire information for you without realising why you require it, or smuggle material unknowingly. Even untrustworthy persons can be used in this way, e.g. black market operators.

iv. <u>Approach to Potential Agents</u>.

    a). <u>Getting to know your man</u>. Never attempt to recruit an agent at a first interview, however highly he has been recommended to you, unless there is no alternative. Always sound your man first and satisfy yourself as to his bona fides and suitability before committing yourself. There are two methods of contacting -
1. By introduction by a mutual friend.
2. Self-introduction.

        1. If the introducer is a first-class man and has your entire confidence, the whole process will take less time, as all you need to do is to get to know your man and form your own opinion as to his suitability.

        2. There are many varieties of self-introduction but the best is the reverse form, i.e. making the man introduce himself to you. (This is the method favored by confidence tricksters.) One way is to ascertain the man's leading inter-

ests and then start a
conversation on the
subject in his pres-
ence with a third
party - e.g. the
bartender. It is
probable that the
man will join in.
Naturally the process
of satisfying your-
self as to his bona
fides and suitability
will be much slower
than in 1.

ORGANIZATION OF AGENTS

GENERAL.

On taking over an area, an organizer may decide to sub-divide in one of the following ways -

a). Geographically.

b). In groups (industrial, political, religious, racial, minorities, etc.)

c). Sub-divide (a) by (b) or (b) by (a).

1. THE HEAD AGENT OR AGENTS.

    a). After a time you will probably be able to select among your agents one or two really reliable men, who can be used as head agents.

    b). Do not put them over men they have worked with. It may cause jealousy. Allow them to choose their own sub-sources and break fresh ground.

    c). Never let head agents know more than

is necessary about your organization and future plans.

d). Remember to keep agents, as far as possible, in water-tight compartments.

e). Make regular checks on even your most reliable agents. You owe this to the security of your organization.

2. <u>CUT-OUTS</u>.

a). Cut-outs are for the protection of the organizer and the agent. Make that point very clear to agents. It is a security measure, and not, as so many agents think, a lack of confidence in their integrity.

b). Method of function. Organizer requires three agents in a certain factory. Instead of contacting them personally, he deals with them through a reliable go-between.

<u>NOTE</u>: It is better for agents not to know the organizer. Even if the agents do know the organizer, it is safer to work through a cut-out.

c). It is advisable, in a largish organization, for the organizer to work each branch or section through a cut-out; e.g. a separate cut-out between railway workers, dockers, factory hands, etc.

<u>Remember</u>: Special care must be taken to keep the meetings between organizer and cut-outs absolutely

secret. Places of rendezvous should be constantly changed.

3. **CELL SYSTEM**.
   This is a system largely used by the communists as a means of infiltrating and permeating the working classes with a view to subversion and, eventually, revolution. A cell is a small compact group working inside an economic or political organization, e.g. a dockyard, a factory, a trade union, a political party. It can be used by us for such purposes as -

   i. Propaganda.

   ii. Political activities against puppet regimes.

   iii. Passive resistance.

   iv. Slowing up production and simple sabotage.

   v. Strikes and labor troubles, (e.g. dockyards).

   vi. Information service.

   vii. Workers going to Germany.

   On the other hand it is useless to form cells without a definite objective.

   a). Cells can consist of any number from three to eight persons. Five is considered to be about the ideal number.

   b). The fundamental precaution of secrecy and security must be faithfully followed. Special regard must be taken to meet and operate as far as possible out of sight of the police, provocateurs, enemy organizations. In factories, ship-yards, railway, etc., from employers and managerial personnel.

c). The essential feature of each cell is that it is as nearly water-tight as it can be made - i.e., only one person in each cell (i.e. organizer) knows anybody in cells above, and only one person (i.e. the liaison man) knows anybody in the cell below.

4. WORKING OF CELL FORMATION.

ORGANIZER

Cell one (Information Agent   Administrative Agent
(
(Agitation, Propaganda,
 Action Agents.

K

<u>LIAISON AGENT</u>
(With Cell 2)

<u>ORGANIZER</u>

Cell   (Information Agent   Administrative Agent
two   (
         (Agitation, Propaganda,
          Action Agents

<u>LIAISON AGENT</u>
(With Cell 3)

5.   <u>FUNCTIONS OF AGENTS</u>.

   Each member of the cell is an active member working towards achieving the objective of cell. (See par. 3). As there are, however, various routine duties which must be performed, these are usually allotted to the various members of the cell, viz. -

   (a)   Administrative duties, which may include collection, storage and dissemination of material, transport, etc.

   (b)   Information, which involves collating all intelligence required to carry out any directives and for self-protection.

   (c)   Liaison. The liaison agent of one cell is responsible for choosing and instructing the organizer of the next cell below. He knows nobody else in that cell. Any member of one cell can be a liaison agent with another cell.

- (d) Members of cells work together and have common aim. They should meet for business purposes as seldom as possible once the general policy has been worked out.

- (e) In order to prevent overlapping (e.g., one worker being recruited by two different cells) organizers of each cell should be given a limited sphere of activity (e.g., one department of a factory, one political branch or a geographical district.)

- (f) Once again, quality before quantity.

6. <u>POLICY AND DIRECTION OF A CELL ORGANIZATION</u>.

    - (a) Policy must be laid down and work controlled by the organizer at the top.

    - (b) The Chief's orders must be obeyed throughout the cell formations.

7. <u>SNOWBALL</u>.

    It will be apparent from the above that it is part of the function of each cell to propagate and throw off daughter cells. Members of the various cells should also mentally collect names of persons who could be used in the event of an armed rising.

8. <u>DEALING WITH OTHER ORGANIZATIONS</u>.

    - (a) Remember, in dealing with other friendly subversive organizations in your territory, that they may have been penetrated by the local S.S., C.E., or police. The use of cut-outs in such contacts is essential.

(b) In countries where there is a legal opposition to the government, the parties in opposition are more likely to be penetrated by government men. If you try to get in touch with them, therefore, your activities are almost certain to come to the notice of the authorities. Use cut-outs again.

9. PROS AND CONS.

The advantages of the cell system are -

(a) Its capacity for penetration into existing organizations, whether industrial, religious, political and even racial.

(b) It affords the maximum security possible in organizations of this kind by -

   i) limiting to a minimum each member's knowledge of the organization.

   ii) in the event of enemy action against one cell it affords time to the others to take the necessary precautions.

(c) Rapid expansive action on the snow-ball principle without equivalent increase in risk.

The disadvantages are -

(a) Lessened speed of action. It obviously takes time to communicate instructions through all the various cells.

(b) Remote control. The leader does not know personally with whom he is dealing or their various capacities.

ESTABLISHMENT AND ORGANIZATION OF H.Q.

1. DEFINITION.

   The H.Q. is the base from which your organization functions. This may range from a highly organized commercial office to a back bedroom on the third floor.

2. ESSENTIALS OF AN H.Q.

   (a) Accessibility. Irregular and special visitors must be able to find H.Q. easily without having to ask for directions from outsiders - e.g. a cellar under a house in a complicated dock-side area back alley is poor from this point of view.

   (b) Cover, for Building. You must be able to account for more numerous visitors than in normal residence. Three types -

      i) Place open to all and sundry - e.g. Barber's shop, grocer's.

         Disadvantage: Complete lack of security control on visitors - e.g. infiltration of agents provocateurs to barber's shop is easy.

ii) Place open to all and sundry, but where visitors may be treated privately and individually - e.g. tailors, dressmakers, doctors, dentists.

<u>Disadvantage</u>: Security control on visitors easier, but by no means infallible.

iii) Place open only to particular type of client, e.g. any kind of office, lawyer's commercial.

<u>Advantage</u>: Difficult for agents provocateurs and enemy C.E. to obtain good reason for visit. Good security control by judicious use of waiting rooms, etc. Possession of safes, etc., accounted for.

(c) <u>Cover for Visitors</u>. Every special visitor must have a "Genuine" reason for his or her visit. Examples -

<u>Tailor's</u>. Visitor comes to try on clothes or have clothes altered. Brings suit with him.

<u>Doctor's</u>. Visitor has complaint ready in case of question.

<u>Lawyer's</u>. Visitor comes on legal business that may have to be carefully "built up" by organizer in advance. Usual office

                    hours should be observed.

    (d)   <u>Camouflage</u>. H.Q. must merge into its surroundings, e.g. the sudden appearance of a smart dress-maker's in a slum district would be poor camouflage. Visitors must be in keeping with the district.

3.   <u>VISITORS</u>.

    (a)   From the beginning instruct actual agents in regular communication with you never to visit H.Q. Receive only specially selected cut-outs and, occasionally, irregular visiting agents.

    (b)   Have the H.Q. running on its "lawful business" for as long as possible before using it for subversive activity. Introduce regularly visiting cut-outs gradually and discreetly, at first, on lawful business only, so that casual outside observers may get used to their visits and view them without suspicion.

        <u>Note</u>: Should a regularly visiting cut-out become "brule", police may inquire from casual outside observers whether he was, in actual fact, a regular visitor.

        From the beginning cut-out may need occasionally and innocently to change his appearance, e.g. visit sometimes in a coat, sometimes in a mackintosh; sometimes in one hat, sometimes in another; sometimes in horn rimmed glasses, sometimes in pince-nez.

He should never, except in acute emergency, adopt active disguises such as moustache, beard or hair-dye. Dark glasses should be used only where customary.

4. <u>ALTERNATIVE H.Q.</u>

At an early state, preferably before your H.Q. is engaged on subversive activity, start a second "spare" H.Q. Do not use this place for any subversive work at any time, until the first is "blown". Let it mature. In extreme emergency it may serve you well as a hide-out.

5. <u>SECURITY MEASURES</u>.

From the beginning pre-arrange the following security measures:-

(a) Pre-arrange a recognition signal in order that special visitors can be recognized.

(b) <u>All Clear signs</u>. To warn special visitors that the coast is clear. These may be:

Curtain drawn, or not drawn, over a window; presence or absence of flowers, etc., in window; red ink on writing desk; re-arrangement of articles outside store, workshop or in shop window.

(c) <u>Danger sign</u>. If, say, a Gestapo agent has openly raised your H.Q. and insists on concealing himself in your office, you may be unable to establish the visible danger sign under surveillance. Arrange, for example, that in normal circumstances a special visitor shall say "Good morning" to you first; but that, in emergency, you will be the first to greet him.

Work on the principle that safety signals should be slightly abnormal, but danger signals completely normal.

(d) <u>Guard</u>. Where your H.Q.'s cover story permits, see that some trusted and responsible person is on duty by day and night. If your H.Q. (e.g. in a shop) is such that there would naturally be no-one on duty at night, so arrange locks of doors, windows, furniture and effects that, if an entry has been made, you will spot it in the morning.

<u>Note</u>: Even if your papers were locked in an unbreakable safe, you would still want to know whether your H.Q. has been entered.

<u>Tricks</u>: Small piece of matchstick balanced on inside of keyhole; small piece of paper on side of drawer. Preferably not breakable cotton-thread obstacles. These are sometimes detectable.

(e) <u>Papers</u>. Keep as few as possible. Memorize as much as possible. In arranging for swift destruction remember that -

    i) Paper, even when petrol-soaked, burns very slowly. Soak in solution of potassium chlorate or interleave every seven pages with celluloid. Beware of betrayal by smoke from burning papers.

    ii) Break up ashes.

(f) <u>Daily Departure</u>. If you suspect that you are under observation, before leaving H.Q. get someone to take a look at the street, innocently, before you go out. Pre-arrange danger signal if he thinks you are under surveillance.

(g) <u>Rents and Taxes of H.Q. Premises</u>. Pay these promptly in order to avoid any inquiry - unless the "cover-owner" of H.Q. has never paid rent promptly before.

L

INTERROGATION

1. **POLICE PROCEDURE**.

    A)   Fresh search prior to interrogation.

    B)   Two man team (or more), one interrogating close to S, other armed near door taking notes; they alternate.

    C)   Placement of S: Not too comfortable - Not permitted to smoke own cigarettes.

    D)   Exterior armed guards.

    E)   Observation Room: Two way observation mirror - recording apparatus.

    F)   Light Signals: Placed behind S - permits communication with Observation Room.

    G)   Laboratory: Documents, clothing, etc. analyzed during interrogation; Couriers for communication and delivery.

2. **PREPARATION**.

    A)   Complete study of dossier: Ready command of all facts vital.

    B)   Psychological Analysis of S: Observe and converse with S <u>prior</u> to interrogation informally to ascertain character temperament and emotional weaknesses - Use of

"Human Relations" expert.

    C) Plan of Strategy: Carefully outline prior to interrogation.

3. TYPES OF INTERROGATION.

    A) Routine:

        No specific objective - Spot checks or infractions of petty regulations. If S natural, posed, consistent, and papers in order, he is released. Any slip refers case to higher authorities or invites more thorough examination.

    B) Clarification: Directed toward clarifying an inconsistency, in papers or a specific point baffling police without any definite suspicion. Short and specific. S. released if he satisfies police. May include test questions of S's cover story, personal life, knowledge of streets and buildings, and recent movements.

    C) The Complete Interrogation: Invoked when S is suspect or fails to satisfy police in (A) or (B) type interrogations.

(1) S's entire life history from cradle with a definite tendency toward the grave.
(2) Complete body and house search accompanies interrogation.
(3) Duration: Days to months.
(4) Coverage: Your lifetime of movements, associations, jobs, skills, relations, opinions.

4. INTERROGATION TECHNIQUES.

A) Irrelevant Opening: May begin with remote facts of early life. S, calculating on crucial issues, is frequently caught unprepared on such matters.

B) "Friend-Enemy" Technique: In two man team, one is friendly and generous toward S, the other badgering and tough.

C) Narrative in line of Q & A: Enables interrogator to ascertain matters S is "willing" to relate and those he wishes to conceal.

D) Scrambling: Disorder, irrelevance, and lack of sequence deliberately employed to disrupt S's orderly preparation; Conceals relevancy from S. Destroys cover story by developing inconsistencies not readily apparent to S.

E) Off The Record: Deliberate device to lull S into sense of false security so that he will engage in social chatter which may reveal important leads.

F) Simulating Mistake or Belief in S's Innocence: Deliberately employed to induce S to believe he is free to operate as usual. Followed by S's release with apologies. Followed by super-surveillance.

G) Pretended Ignorance or Misstatements: Intended to make S. believe interrogator unaware or misinformed so that S will try to capitalize on the situation by affirming a fact known to be false by the interrogator. (Cf. S must have observed fire at the hotel he stopped at last month).

H) Introduction of Humane "Higher Up": Deliberate device to induce repetition of S's testimony under a more humane interrogation from a more "reasonable" and authoritative official who is scornful of uncouth bungling by petty police.

I) Apparent Failure To Transcribe Testimony: No notes taken by anyone so S thinks he is undergoing informal inquiry. Recording machines in operation.

J) Confrontation With Forged Confessions or Admissions of Associates: Faked recordings and forged confessions. Or true recordings out of context.

K) The Surprise Visit: Your door battered down in middle of night invasion of flashlights - yanked from bed and interrogated when defenses are down and S wholly unprepared.

L) Suggesting Duplicity: Fictitious proof to convince S that he was "sold out" by his associates or that he is the "fall guy" for superiors in his organization.

M) Exaggeration by Interrogator: Of extent of S's participation or importance in certain activities to arouse S to vehement argument during which he may make admissions.

N) Friendly Fellow Prisoners, Nurses and Guards: Really enemy agents subtly winning S's confidence thru cautious expression of anti-fascist sentiments.

O) Offers of Mitigation: Alluring hope of release, "exchange", etc. Frequently backed by forged evidence of high authority for such action in S's case.

P) Silence Treatment: Alternating severe grilling with long periods of solitary confinement. Or during grilling long periods of silence while interrogator examines dossier producing discomfort and unsolicited statements.

Q) Injury or harassment of Relatives or Friends: Importance of ascertaining facts re relatives and friends and knowing extent of S's emotional attachment to them. Use of family hostages a Nazi specialty.

R) The Third Degree: Nazis have combined sadism and science. The Dentist's Lab and Big Bertha in Sofia. More common: (1) Rubber hose (2) Kleig lights day and night and S sees no one. (3) Offering incriminating papers to sign after resistance reaches breaking point - confrontation later when S revived. (5) long grilling followed by solitary, rats and vermin. (6) Forcing S to observe torture or execution of others.

    Limitation: Enemy knows that after some testing that torture will be ineffective on certain individuals; also danger of insanity or death which kills the goose who may lay an informative egg some day.

5. <u>MECHANICAL AND PSYCHOLOGICAL DEVICES</u>.

   A) Polygraph (Lie Detector): Records breathing, heart beat, blood pressure.

   B) Scopolamine (Twilight Sleep): Like extreme intoxication. Induces loose, uninhibited talk.

   C) Hypnotism and Post Hypnotic Suggestion: Difficult - especially if repugnant to deep seated moral or patriotic convictions.

D) Free Association Test: Numerous "key words" interspersed among far greater number of innocent words. Stop watch records time of spontaneous responses; unusual length or brevity of time for response plus degree of irrelevancy indicative of falsity. Sometimes effective. Can develop a defense by rhythmic, normally long response interval.

PERSONAL DESCRIPTION & IDENTIFICATION

1. CHIEF USES IN OUR WORK.

   A) To recognize an unknown person by description only. (CF. enemy personnel or unknown friendly confederates.)

   B) To describe to your confederates or home base a person unknown to them.

   C) To elicit an accurate description from persons furnishing the information unwittingly. (Cf. landlady, bartender, maid, etc.)

   D) To establish the identity of persons responsible for certain activities or statements. (Cf. public officials, industrial leaders, key people in enemy and friendly groups.)

2. SCIENTIFIC DESCRIPTIONS.

   A) Anthropometry: Bertillen's system of body, head and limb measurements in 65 divisions and over 65,000 classifications.
   B) Portrait Parle:

        (1)   Color (eye, hair, beard, and skin)
        (2)   Morphological (shape, size, direction of parts of the head)
        (3)   General (build, carriage, voice & language).
        (4)   Indelible Marks (scars, tattoos, etc.)

    C)   Fingerprints: Galton-Henry System.

        (1)   10 finger classification.
        (2)   Single print classification only of prints found on scene of crime.
        (3)   Plastic Prints: In soft, plastic materials (soap, tar, etc.)
        (4)   Visible Prints: In colored materials (blood, ink, dirt).
        (5)   Latent Prints: Invisible until revealed (oblique light, breathing on object, developing)
        (6)   Palmar and Sole Prints.

    D)   Photography: (1) Full face and profile.
                       2) Reveals ear characteristics.

3. <u>TAKING DESCRIPTIONS</u>. The "Approaching Subject" methods best for our type of field work where one must rely entirely upon memory.

    Imagine the Subject approaching you from a distance:

        A)   General: First thing you note is height, next build, then

		gait, then age, social status.
B)	Full Face:	As subject approaches closer, details of face become apparent. Begin at top observing hair, head shape, forehead, eyebrows, eyes, nose, mouth, chin, neck, markings (scars, dimples, clefts, etc.)
C)	Side Face:	As subject passes, he is in profile. Observe features in same sequence with special attention to the ear.
D)	Clothing:	When you intend to furnish description to someone who must identify. Subject when he is likely to be wearing same clothing, the clothing description is of paramount importance and should be observed after (A). If description for home base or not for immediate use, then last.

4. <u>GIVING DESCRIPTIONS</u>.

   Following order easier to categorize in your memory and easier to recall:

   A)   General.
   B)   Clothing.
   C)   Top to bottom.
   D)   Habits, haunts and associates.

5. <u>SPECIAL FACTORS IN DESCRIPTION</u>.

A)   Height:

     5'   to   5'3" .... small
     5'4" to   5'7" .... medium
     5'8" or   over .... tall

B)   Baldness:

     Frontal, Occipital, Top or Total.

C)   Eyes:

     Bloodshot, Watery, Pear shaped,
     White spots, Lashes, Bags and Folds.

D)   Eyebrows:

     Color if different from hair; bushy,
     continuing across top of nose.

E)   Nose:

     Saddle wide or narrow; Bridge broken, flat, deviating left or right; Nostrils thick, thin, or strongly revealed; Point upturned, blunt, or split.

F)   Ears:

     Upper helix pointed or angled; Lobule adhering to cheek, pierced, or split; or twisted; Position close to head or protruding; General - cauliflower, huge and tiny.

6. PRACTICAL DEMONSTRATIONS.

    A)   Describe various instructors not present from memory.
    B)   During lecture enter stranger of incongruous appearance, vivid cloth-

ing, bearing various objects. He addresses previously prepared remarks to instructor (addresses instructor by incorrect name, gives list of items to be purchased, etc.) On way out, he drops some object. Lecture continues. After five minutes, students asked to note on paper personal description, including clothing, actions and oral statements of stranger. Correct papers for inaccuracies and also note item omitted.

DESCRIPTION AND IDENTITY PARADES

Object: To utilize the knowledge afforded by Lecture on "Descriptions".

A) DESCRIBING A MAN.

In this exercise a living model will be placed in front of the class or photographs will be thrown on the screen by the opidiascope. Students will be asked to recognise and name salient features and afterwards to write down descriptions of the individual or photograph. They may also be asked to identify a man's photograph from a description.

B) IDENTITY PARADE.

1. Students will be given the description of a man, preferably not more than four salient features, two of which should normally be his height and apparent age. Students will then proceed one at a time to an identification

parade consisting of as many men as possible and will be given thirty seconds in which to identify the man.

2. Each student will stand beside a door or gate and a parade of men will come out one by one. Here again, Students will have to identify the man from the description in their possession.

3. In the descriptions given in (1) and (2) one strong salient feature can be omitted by way of change and the Students warned of this beforehand and asked to state, after they have found their man, what is the feature omitted.

M

BURGLARY

1. DEFINITION:

   Breaking and entering at night with intent to commit a felony. Lesser degrees of crime where "night" and "felony" omitted.

2. BURGLARY BY TRICK:

   A) Simulating ownership or right: (Cf. Dillinger bank robberies simulating movie shots; stealing telephones dressed as repairman; Major ....'s chinaman stealing clock from Hong Kong courtroom)

   B) Entry by Ruse: (cf. posing as plumber, fake telegram, fire, etc.)

   C) The converse: Apprehension through appearance of guilt (cf. removing park bench which you actually own; running when hue and cry raised)

3. FORCIBLE ENTRY IN ABSENCE OF BURGLARY ALARMS:

   A) Forcing or Picking the Lock: (see lock picking memo) Shooting the lock; foot pressure against the door then suddenly striking the door at the lock position.

   B) Remove Door Panels

   C) Windows:
      (1) Double hung: knife, jimmy, or slipper to move collar; if bolted, drill hole to end of

                  bolt and punch bolt out.
- (2) French type: Remove hinges; or remove pane adjoining lock and operate lock manually.
- (3) Removing glass: (a) Cutting: too noisy (b) Tape pane with friction, adhesive, or scotch tape - flypaper or other adhesive paper will do - break sharply blending with external noise (c) By trick; remove transom or other obscure pane under cover of handyman during day in preparation for night entry.

    D) Barred Windows: Usually psychological barrier only when bolted to outside masonry. Remove bolts with Stilson wrench. If in masonry, pry apart with auto jack. File notches in bar to catch jack at an angle if bars too close together.

4. <u>FORCIBLE ENTRY DESPITE BURGLARY ALARMS</u>:

    Modus Operandi trademarks the professional. They specialize in particular crimes and develop an individual technique which bears the signature of the intruder. Hence police departmentalize (Safe and Loft Squad, Alien Squad, Narcotic Squad, Frauds Division, etc.) <u>Police Classification System</u>: By numbers on cards, each representing an element of the crime (cf. nature of crime, means of entry, technique of entry, type of premises or vehicle, special trade marks, etc.

    A) Roof Burglar: Entry through roof via fire escapes, windows of adjoining buildings, standpipes, cut power or telephone lines, rope ladders, etc.

B) Wall cutter: Rents adjoining premises and saws or bores through wall to premises to be robbed. Usually on week-ends.

C) Tunnel Burglar: Rents adjoining or neighbouring premises and tunnels through. Raises concrete floors of premises to be burglarized with heavy auto jack or moving jack. Objective usually well protected warehouses. Sometimes planned for months.

D) Hideout Burglar: Hides in building after closing hours; in toilets, lockers, top of elevator, etc.

5. <u>TYPES OF JOBS</u>:

   A) Safe and Cabinet Jobs:

   (1) By manipulation: With use of Code books and skill of combination experts. Rare.

   (2) Rip Jobs: Use sectional jimmy, five foot "can opener", crowbars, drills, sledge hammer, pick. Doors of poorly secured safes have thin metal cover with brick or cement inside. Drill hole in upper left hand corner and rip steel covering with "can opener" then pick out brick and cement; work combination lock from inside. Cannot drill modern manganese steel safes.

   (3) Punch or Knob Knocking Jobs: Sledge Hammer knocks out dial

- 247 -

and shaft punched back with a center punch.

(4) Drag Jobs: Pulling out dial and combination spindle by pressure of screws tightening against a V-shaped slot placed under dial. Tedious and difficult if pressure not evenly distributed.

(5) Chopping Jobs: Turn old type safes upside down and chop bottom out. Impossible on modern safes where bottom well secured.

(6) Blow Jobs: Nitroglycerine poured into soap or bread dough mould placed along crack of door; flows into safe. Dangerous to handle! Requires just enough to blow door without damaging contents; or drill hole left and above dial, fill with cotton soaked in nitroglycerine wrapped around fulminate cap. Blankets or sacks used to kill sound which is timed to blend with noise of passing truck. In modern safes crack paper thin but hypo needle under pressure used to shoot thinned nitroglycerine into crevices.

(7) Torch Jobs: Only good on cabinets and small safes because acetylene tanks too bulky. Burn out combination. Danger of burning contents. "Floater" fill first with a

water to protect money and papers from heat. Modern safes have bronze and copper layers to carry off heat quickly and prevent consistent melting temperature.

(8) Sleeper Jobs: Where dial not fully twirled to release tumblers, turn back and try bolt handle. Janitors and charwomen employed by agents to test all safes for weeks until they catch a sleeper. Victim never detects the entry. Silver nitrate powder may be used to catch inside thief by tell-tale black fingers.

6. <u>PLANNING THE ESCAPE</u>: Plan <u>Prior</u> to entry.

   A) Gimlet, screw, or nail inserted into frame at lock position. If owner returns, he thinks lock jammed.

   B) Duck behind door when opened - as person enters, slip out or conk him with jimmy if he turns around - then continue with the job.

   C) Emergency Exits: Windows, vines, previously cut telephone or power lines, ropes, etc.

   D) Escape by Tricks in Hotels, pretending you are in wrong room - open suitcase, etc. Makes you conspicuous.

7. PETTY THIEVERY:

    A) Pickpockets:

        (1) The Jostler: Magician's technique of diverting attention.

        (2) The Dip: Sudden quick movement - use of both hands preferable. Letter Snatching.

        (3) Groups: One or more jostlers (open buttons), dips, fronts (to receive passes), barkers to warn of pickpockets in order to locate wallets. (cf. Practice in dummies a la Fagin)

        (4) Paper Lifting: Throwing newspaper or hat over document during conversation. Upon leaving "accidentally" lifting document. (countermeasure: Planting documents containing doctored information for the very purpose of having them lifted.) When lifting documents, take other papers in addition to those desired to spread suspicion and hide object of theft.

8. THE ORGANIZED INTELLIGENCE RAID:

    A) Entry: Connivance of owner, by trick, lockpicking, or forcible.

    B) Reconnaissance: To determine guard schedule and hideout for all night job.

- C) Control of Elevator: To prevent surprise return.
- D) Sentries: Interior and Exterior.
- E) Radio Communication: Between radio car in street and job.
- F) Lock and Safe Experts.
- G) Evaluators: Intelligence (to determine importance) and Language Evaluators (to translate).
- H) Automatic or Bellows Photography: 500 to 1000 per hour Photograph entire code books page by page.
- I) Dust atomizers: Conceals all traces of entry.
- J) Couriers

  (cf. Feb. 1941 - Kidnapping of Greevitch, British Passport control officer at Sofia on Orient Express to Istanbul. Raids on British Consulates in Burgas and Sofia while England neutral).

9. <u>FLAPS AND SEALS</u>:

- A) Flaps: Dry opening and steaming outmoded.

  <u>Wetting method</u>: Formula -

  1 gram FK Water Spot Preventative
  8 ounces distilled water
  15 drops Laboratory Aerosol - 10% solution
  3 drops of clorox

Wet, open flap, replace glue not too close to edge, iron flat.

B) SEALS:

(1) Use Albastone or Coecal for reproducing the seals.

(2) Cover entire wrapper with cellophane or other transparent covering, cutting out outline of seals so as to reveal the seals for work and prevent damage to wrapper.

(3) Apply 10% solution of Laboratory Aerosol with brush (or a light oil)

(4) Mix Albastone or Coecal moulage with 5 drops of 10% Aerosol and 1 ounce distilled water. Spatulate until dissolved. Thick, creamy mixture.

(5) Spoon a little to seal: set 2 minutes, add more and build up to 1 inch cone. Set 10 minutes until firm. Pencil mark position; then remove and trim.

(6) Clean seal with scalpel and water.

(7) With warm soldering pencil, cut seal at flap line (not too hot to melt); do not cut to paper O break balance.

(8) By wetting method upon flap.

(9)　　Replace flap; weld crack in seal; then melt until creamy.

(10)　Oil reproduction copiously; affix to molten seal; remove when set.

BURGLARY

In the limited time at our disposal we will attempt to cover some of the most common methods used by burglars in their various fields not necessarily for the purpose of becoming adept as burglars ourselves, but in view of the fact that we all are adopting the viewpoint of a foreign undercover agent it is certainly not inconceivable that the methods about to be discussed may be used against us by enemy agents and in order to successfully neutralize these attempts it is well to anticipate their actions and approaches beforehand. Going still further, many of the techniques which will be brought out may stand us in good stead at some future date.

At this point it should be noted that the techniques of burglary which are considered reprehensible in peacetime are actually the practical application of methods used in modern warfare to attain a given objective. As civilians you are undoubtedly familiar with most of the terms which will be used due to the wide publicity given to criminals and their operations by the press. The types of crimes committed will also be quite familiar and the actual method of accomplishing the acts will be found in most instances to involve the application of physical principles and psychological approaches which are unconsciously used almost daily by all of us in a law-abiding capacity.

Since the techniques to be covered are not sufficiently distinctive to term burglary a "science" it might be stated that the only thing which distinguishes a burglar from a law-abiding citizen is his attitude toward society, his failure to recognize the personal and property rights of others, and the fact that he uses other than legal means of attaining his ends.

The purpose, then, of our discussion is to place ourselves in the position of the burglar, adopt his frame of mind, and train ourselves to look for the objectives with which he would concern himself. Having once established our objectives we will apply principles with which we are all familiar to their attainment.

For purposes of discussion we will informally cover various types of crimes, noting particular methods used in specific cases when possible, and if a general method can be shown this will be done.

It should be realized that this discussion does not purport to thoroughly cover the entire field of crime or to go into any particular case with any degree of thoroughness. Such a complete discussion would obviously be limited only by the number of specific examples of crimes available.

Possibly the most important thing to remember in the commission of any crime is that you yourself (the criminal) are the only person who knows what is going on and that when the police arrive on the scene of the crime they know absolutely nothing and must start to work with the evidence which immediately is available. A very large percentage of criminals have been apprehended on the scene of the crime or nearby by reason of the fact that they attempted to run, thereby making themselves conspicuous. It is a cardinal rule, therefore, that one must be calm and keep his wits about him at all times. For example in the City of New York a person will probably be picked up by a policeman merely by reason of the fact that he is running down the street. The policeman will then retrace with the person a distance of possibly one or two city blocks just on the theory that perhaps some crime such as pickpocketing or shoplifting has been committed. A person in such a case would have made himself conspicuous and thereby defeated his own ends. A good example of the application of this principle occurred some months ago in New York where a particular gang of robbers had successfully staged armed holdups of various saloons throughout the City over a period of several months and despite the fact that none of the group ever wore masks and there were always plenty of witnesses it seemed impossible to ever identify any of them and the police were totally unable to obtain any lead as to whom might be committing the outrages or as to how they could so completely vanish within a matter of a minute or two after the commission of the robbery. The gang was eventually apprehended and it developed that on the occasion of every robbery they would caution the victims not to call the police for at least five minutes after they had left, threatening dire results, they would then proceed to run to the nearest alleyway, change hats and coats, and return immediately to the scene of the crime, to all intents and purposes merely innocent bystanders attracted by the hue and cry of the victims. In no case did any of the victims ever recognize any of the criminals among the bystanders, and the group was finally apprehended only by reason of the fact that one of the members became nervous, lost his head so to speak, and attempted to escape from the scene of the crime by running and was thus picked up by the police merely because he had made himself conspicuous and was doing what would have been expected of any criminal under the same circumstances. This same example is also a good illustration of the fact that the average person under the strain of a robbery is unable to give a reasonable description of the criminals and there will probably be as many different descriptions as there are victims. Another good example occurred several years ago when the late

John Dillinger was in his prime. Dillinger and his gang had decided to stage an armed robbery of a bank in Sioux Falls, South Dakota, and accordingly on the day picked for the robbery they appeared on the scene in an automobile. Dillinger and several members entered the bank and announced that this was a holdup, and at the same time several members of the gang were stationed on the various street corners adjacent to the bank to insure that there would be no interruption during the course of the robbery. The men were openly armed and were directing the crowd that gathered in much the same manner as a policeman directing traffic. A very large crowd gathered and the story is told of the kind old lady who stepped up to the gunman and asked what was going on and he very calmly pointed over to the crowd and told her she had better go over and stand with the rest of the people, that it was merely a bank robbery taking place. An employee of the telephone company drove unwittingly onto the scene of the crime and was instructed by the gunman as to where to park his car and to join the rest of the crowd. The spirit of the occasion seemed to be one of interest on part of the crowd and it was not fully realized by them that the affair was deadly serious, until the robbery was completed. The people in this instance had never witnessed such an armed robbery before and their minds refused to believe that such an event would be carried on in so obvious and straightforward a manner, the very audacity of the act accomplishing the desired result.

It frequently becomes necessary to break down doors either for purposes of committing burglary or from the standpoint of the officer to apprehend a criminal. In this connection the approved movie style which consists of running at the door from a distance of ten feet or so and throwing your shoulder it is very unsatisfactory as you are very liable to wrench your back in the process. If the door to be opened happens to be in a narrow hallway you can, with the aid of another person, very easily open it by bracing your back against the wall, place one foot near the lock, then press on the door until it fairly breaks. At this point have the second person deliver a sharp, hard kick to the door also near the lock and the lock will shoot out of the bolt, frequently shooting entirely across the room. Another practical method was described by a narcotic agent who had injured his back in an attempt to break a door by rushing it, and his theory was that it was much easier to have the people who were inside just to open it themselves. In dealing with negroes who would gather in a room to smoke drugs or "dope" his favorite method of accomplishing this was to approach the window of the room by fire escape or some other means, pull out a big shiny pistol, break the window glass and wave the pistol around inside the room. There would be a mad rush for the door by the negroes and at this point it was always arranged to have someone kneeling just outside the door and the result can easily be imagined.

     In connection with doors, it may be necessary at some time to conduct a private search of the room. In this case after having first conducted a surveillance of the place and ascertained the hours during which occupant is absent so as to give yourself sufficient time to conduct your search without fear of interruption and having obtained entrance into the room by either picking the lock or some other method, your first consideration should be to establish some means of exit through a rear window or door, or perhaps a fire-escape. In any event you would be in a position to leave hurriedly and quietly should the occupant return unexpectedly. However, if there is any possibility of interruption, you may insure privacy by taking a gimlet, which is a slender metal rod threaded at one end, and screwing it at an angle through the edge of the door and into the door frame thereby screwing the door shut. Should the tenant return he will try to lock without success and will probably leave in search of the house manager to complain of the lock which is evidently faulty. In the meantime you have removed the gimlet and quietly gone about your business and it is improbable that the small hole made in the door will ever be noticed and by the time it is the average person would be at a loss to account for it. Should you by some chance be surprised by a returning tenant and the door is not secured and you have no other method of exit just remain calm, always remembering that you are in control of the situation and that the occupant knows nothing of what is going on. In this case it is very easy to step to one side of the door and next to the wall just as the subject is entering and a slight tap on the back of the head will take him out of the picture. A person entering a room is entirely helpless when attacked from this position which is a "dead" "spot" and especially when they are not anticipating something out of the way. To illustrate what an awkward position this is a case in which several federal agents were attempting to arrest a known desperado whom they knew to be in a certain room may be cited. The agents approached the room and ordered the wanted man to come out and he thereupon invited them to come in and get him. One of the agents then attempted to rush into the room but upon stepping inside the door he discovered the criminal standing immediately to one side of the door with a gun pointed at him and uttering dire threats to the effect that he was going to very shortly if not sooner shoot the particular agent full of holes. The situation was a stalemate as the one agent could not move without being killed and the other agents

still in the hallway could not give him aid for
fear of this same thing and the criminal knew
that should he kill the one agent then he himself would be killed by the others. The situation finally was solved by the one agent arguing
with the would-be killer until he had literally
"fast talked" himself out of the situation. If
entrance into the main entrance of large apartment houses is desired it is a quite simple matter to wait until someone comes out of the door
and then immediately step in while the door is
still open. This action is quite normal and you
could very easily be another tenant in the
building so far as someone else is concerned.

A "jimmy" may also be used to open doors,
windows, cabinets, etc. This instrument is simply what is more commonly known as a "crow-bar"
or in other words a steel bar about an inch in
diameter and probably three feet in length with
one end pointed and the other end flat so as to
be useful in prying open any two objects. A
professional "jimmy" consists of two sections
which are fitted together with a metal collar
and the entire unit fits very nicely into a
leather carrying case about a foot and a half in
length. In lieu of either a "crow-bar" or
"jimmy" a very handy substitute may be found in
the form of a metal brakeshoe key used on all
railroad freight cars and since many freight
cars in foreign countries are of American manufacture this would also be a good source of
tools if abroad. The "key" itself is a slightly
curved steel bar possibly fourteen inches or
more in length and is used to hold the brake
shoe in place on the freight car. The steel is
not as well tempered as that of a professional
tool but it is very satisfactory for most purposes.

Another method of gaining entrance is
through a glass window or skylight. This can be

done by removing the putty which holds the glass in place with a knife or other similar instrument, removing the small metal pieces driven into the window frame to provide mechanical rigidity and simply lifting the glass pane out. If the pane is sufficiently large entrance may be made directly through the opening thus made. On the other hand it will be noted that many doors have panes of glass directly to one side of the door lock and handle. In this case merely remove one pane of glass, reach inside and open the door. If it were desired the pane of glass could be replaced in the frame and could possibly be set in place with fresh putty. Another more direct method of getting the glass out is to break it out. In this case of breaking glass the noise is caused almost entirely by the falling glass and not by the breaking. It is an easy matter to cover the glass pane with adhesive tape or other material of adhesive character and then break the glass by a short blow with a blunt instrument, if necessary this could be done with the gloved fist taking care not to run the fist or instrument entirely through the window. The result will be that the glass will shatter and adhere to the adhesive and can then be lifted out of the frame. With care this operation can be accomplished with a minimum of noise and will attract no attention whatsoever under ordinary circumstances.

In the matter of forcing entry through barred windows it will be noted that in many instances, especially in small towns, the bars across the window with the bolts on the outside. Obviously the only instrument required is a wrench to remove the bolts and the bars can all be lifted off. Should this not be the case it is well to keep in mind the possibilities of an automobile jack, especially one of the hydraulic type. By inserting the jack between the bars and operating the lever the bars will be easily bent apart or pulled out of their sockets. The story is told of the salesman who dealt in burglar alarm devices and whose favorite customers were small town banks. His routine would be to first approach the bank president with the idea of installing a burglar alarm device. Almost invariably the response would be negative since the bank was sufficiently protected with bars on the doors and windows. The salesman would not press the point but during the night would go down to the bank, place an auto jack between the bars on a window and jack them well apart, leaving the jack in between the bars. A visit to the bank president on the following day would almost always result in a sale of an alarm device. A jack could be used in any situation which required the application of extreme pressure and its possibilities in opening doors, cabinets, etc., should be kept in mind.

In the matter of opening safes there are generally two methods used by professional safe crackers. The first is by the use of an acetylene torch in which case entrance is merely always made through the bottom of the safe, this being the most vulnerable part. In this method it is of course necessary to have a supply of water which is used to cool the metal and prevent the destruction of documents and paper money inside the safe. The other method is to blow the door off the safe by the use of nitroglycerine or "soup".

In this method the crack around the edge of the safe door is filled with yellow soap except for a portion at the top into which a small hole is drilled and filled with the proper amount of liquid nitroglycerine. The safe is then covered with burlap bags to muffle the noise and the charge is set off with a length of fuse. If a good professional job is done the door will come off it hinges and fall at the foot of the safe and the resulting muffled sound cannot be heard fifty yards away. There have been instances in which such jobs were attempted by persons who were not professionals and in at least one instance it resulted not only in the door being blown off but being blown through the side of the building and across the street and silver currency being scattered all over town all as a result of an overcharge of explosive. In using such methods the robberies themselves generally were carried out during the early hours of the morning for the rather obvious reasons that there were no people about and it gave the criminals more time in which to do their job. This weakness in bank protection systems resulted in the introduction of the time lock on bank safes in which system the safe itself is set to open at a predetermined time and until that time it cannot be opened even by the person who set the time device. Ordinarily then the safe would be set to open at or around nine o'clock in the morning at the beginning of business. The criminals also changed their tactics accordingly and adopted the simple method of just entering the bank shortly after it had opened in the morning and just remaining until the safe did open on the due course of time, meanwhile keeping all customers and employees at gun point.

    A favorite pastime in the underworld, especially in the large cities is that of "fishing". As the name implies, this consists of using a long fishing pole with a hook attached, and by this means a man's trousers or other articles of clothing which might be lying about the room on a chair could be picked up and lifted out a window and the wallet removed.

    The art of the pickpocket has also been developed to a remarkable degree and some idea of what may be accomplished in this line may be obtained from the performances of professional magicians who also demonstrate the methods of picking a man's pocket before an audience. Some of these professionals are so accomplished that they can practically take a man's suspenders off without him being aware of it. Professional pickpockets are rarely if ever so proficient yet it is almost impossible to tell when you are being robbed. One method used is to have one member of the gang operating a small street corner concession of some kind. When a sufficiently large crowd has been attracted by his sales talk he will warn the crowd to beware of pickpockets. Almost unconsciously persons having large sums of money will feel for their wallets thus disclosing not only where the wallet is but the fact that they have a considerable amount of money on their person. It is then only a matter of minutes until the gang has completed its work and the operator closes his concession and moves on. Another method is to follow the victim until he gets in a dense crowd or a crowded streetcar or a subway at which time one of the gang will give him the "jostle", that is, will ostensibly bump into the victim accidentally at which time he will unbutton one or more of the buttons of the victim's coat or vest.

    If necessary another member of the gang will then repeat this process and entirely open the victim's vest or coat at which time there will be another jostle by a member of the gang at which time still another member will simultaneously reach inside the victim's coat and extract the wallet. The member who actually gets the wallet will immediately pass it along to another person who is working with the gang but who has no criminal records and who is on the surface a respected citizen. This is done because the local detectives are generally acquainted with all the professional pickpockets and are liable to pick them up on sight just on suspicion, and it is too risky to carry "hot" or stolen goods on their persons. Not only do the pickpockets actually go down into a person's pocket to get at his money or wallet but by the use of small razor-like knives they will cut a man's clothing from the outside and cut out the entire pocket containing the money, making away with money, pocket and all. In these cases a person has a good chance of getting his wallet returned but of course, minus the money. A low form of pickpocket is known as the "lush diver" or in other words a person who steals from drunks who are too helpless to help themselves. An amusing incident occurred in which one of the so-called "Lush-Divers", who was also an informant for the police, observed two other members of his profession "rolling" a drunk, or in other words stealing his wallet. The observer immediately accosted the two thieves and demanded that they give him a share of the "take" or he would turn them in to the police. The thieves were very obliging and gave him 10 dollars as his percent- age. It developed later that the wallet had con- tained a considerable amount of money and the thieves had in reality given the observer only "chicken-feed" and nothing near the percentage he had expected. The unlucky observer was later unburdened himself to the federal agent for whom he worked as an informant and loudly bemoaned the fact that a person just couldn't trust anyone anymore although he himself would have done the same thing had the positions been reversed.

It may be briefly mentioned that second story men among other things use a rope ladder constructed of light quarter inch rope for the purpose of scaling picket fences or any barrier with sharp projections at the top. Some of these ladders are very ingeniously constructed of a single length of rope so woven as to form regular steps and the method of using is simply to throw the top portion of the ladder over the top of the picket fence where it catches and they climb right on up.

In conclusion it should be pointed out that every professional criminal regardless of his field of activity, has his own peculiarities in the type of work which he does and his methods of going about his work which tend to mark a given job as having been done by him just as surely as if he had left his fingerprints with the police. Perhaps the only trademark is that of a job smoothly done with a professional touch which sets it apart from the work of an amateur in a manner in which the criminal can never overcome as they take a pride in doing a good job just as any master craftsman. On the other hand some criminals are positively eccentric in that they take pride in leaving some mark which identifies the work and sets it apart from that of any other criminal.